PORSCHE
The 4-Cylinder, 4-Cam Sports & Racing Cars

By Jerry Sloniger

Motorbooks International
Publishers & Wholesalers

To Prof. Ernst Fuhrmann, father, godfather and dutch uncle to a remarkable engine.

This edition first published in 1993 by Motorbooks International Publishers & Wholesalers, PO Box 2, 729 Prospect Avenue, Osceola, WI 54020 USA

© Dean Batchelor Publications, 1977, 1993

This is a reissue of the 1977 original edition with corrections and revisions

All rights reserved. With the exception of quoting brief passages for the purposes of review no part of this publication may be reproduced without prior written permission from the Publisher

Motorbooks International is a certified trademark, registered with the United States Patent Office

The information in this book is true and complete to the best of our knowledge. All recommendations are made without any guarantee on the part of the author or Publisher, who also disclaim any liability incurred in connection with the use of this data or specific details

We recognize that some words, model names and designations, for example, mentioned herein are the property of the trademark holder. We use them for identification purposes only. This is not an official publication

Motorbooks International books are also available at discounts in bulk quantity for industrial or sales-promotional use. For details write to Special Sales Manager at the Publisher's address

Library of Congress Cataloging-in-Publication Data
Sloniger, Jerry.
    Porsche, the four-cylinder, four-cam sports & racing cars / Jerry Sloniger.
      p.   cm.
    Originally published: Porsche, 4-cylinder, 4-cam sports, and racing cars. Dean Batchelor Publications, 1977.
    ISBN 0-87938-707-6
    1. Porsche automobile—History.   2. Automobiles, Racing—History.   I. Title.
TL215.P75S52   1993
629.228—dc20   92-33867

On the front cover: Porsche 904. Although this is a facsimile edition, the photographs on the cover have been replaced with other photos of the cars originally shown as the film for the original cover was lost. *Chuck Queener*

On the back cover: Porsche 356B GTL Abarth Carrera. *Jerry Sloniger*

Printed and bound in the United States of America

PORSCHE The 4-cylinder, 4-cam Sports & Racing Cars

CONTENTS

Foreword
Engine Development
550 Spyder
550A/1500RS
718/RSK
RS60
RS61
F2/1
Carrera
Carrera Abarth GTL
2000 GS/GT
904 GTS
Specification Charts
Specials

Foreword

In an automobile age noted for model diversity a firm like Porsche which flourished on constancy seems the rebel. Stuttgart's designers stood out for their single-minded development of only two or three basic engines, but they lifted those to extremely high peaks.

And the most self-willed of all their engines in a full quarter-century of Porsches was that four-camshaft, four-cylinder, air cooled, opposed "boxer" which propelled both racing Spyders and plush road cars. It even put Porsche onto the Grand Prix grids, and earned a reputation as the engine that wouldn't quit; a giant killer. It was also well known as one of the all-time horrors to adjust and service.

Of course it is equally true that old Porsche hands saw the 550, 356, F2/1 and 904 projects, all powered by this engine, as very different cars indeed. They point out that Porsche actually built the one famous engine—with two different main bearing layouts—in at least five basic sizes plus several other capacities for one-off tasks.

This powerplant which carried works project number 547 (later 692 and 587 as well) came into being because their ex-Volkswagen pushrod prime mover was at the end of its career for high speed cars.

As Dr. Ernst Fuhrmann—who went from engineer to technical director and then board chairman at Porsche—put it: "We were proud of our alterations to the old engine but its possibilities for housings, cooling and passages were all exhausted. So we debated our first totally-Porsche engine.

"You can only take a pushrod engine to perhaps 6500 rpm before it becomes critical. Even then you need enormous valve overlap. It will be a bad engine. So the thing I really had in mind was not a specific new engine but rather the failings of its predecessor. How could we improve on that?"

The use of "I" here is more justified than it might be for any other engine built since World War II—if not for any engine in this century. In an age of team efforts one man designed, built and tested the four-cam Porsche.

In Fuhrmann's words, "It was the first complete engine I ever did. Also the last. I was very thankful to have such a chance. Before that I had no special field—just theory, stress analysis, whatever was required. But since I have spent most of my working life with engines now, I probably understand them better than chassis."

That life started with the four-cam in 1952—although he had been with the firm since their Cisitalia project, another engine noted for complexity and high promise.

As for 547; "there were no assistants at first. I did all the drawings alone. Normally it would have gone on to a development department then. But when I finished they said, 'we could never get parts to build anything like that.' So I got into my own car and visited the suppliers, talking them into delivering my parts.

"When these arrived they said, 'well, the parts got here all right but nobody could ever put such an engine together. It is far too complicated.' So I went to the test department and found two really outstanding mechanics [Trostmann and Stortz] and we built it but then I went into the test department too and worked on the dynamometer.

"I was standing at Freiburg hillclimb when a camshaft bearing broke in the first race too. Nothing critical. It was under-dimensioned and cast alloy. Later we used forgings. And I was there for the first successful race as well."

There were so many racing wins over such a long span that the design must have been right.

"I had a lot of luck," is the way Dr. Fuhrmann puts it. "The principle was healthy and it was what I call thermally symmetrical. I would credit that one feature for most of the success which really came from the engine's durability." He goes on to explain that this heat expansion factor, almost overlooked in the heyday of the four-cam racers, meant that each pair of cylinders could expand outward from the central kingshaft (layshaft) drive between them at the same rate. It was actually a bonus stemming from a desire Fuhrmann had from the first but seldom mentioned.

"The engine was only planned for racing in a Spyder. But I was a keen driver then too and had secret thoughts about putting it into our 356 as well. Thus it couldn't be any longer than the pushrod engine. And we had to use overhead camshafts of course. But I couldn't fit in the train of gears or chain which usually drove cams in those days.

"So I settled on kingshafts running between each pair of cylinders. They took up little space. With air cooling it also meant that each pair of cylinders could expand outwards without distorting the valve drive." Thus the engine received its most arresting feature, the complicated but accurate drive to four overhead cams.

Thanks to his desire for a faster personal car, Fuhrmann produced a race winner beyond even Porsche's own dreams.

Twenty years later Dr. Fuhrmann added, "If I were given the same task now—fitting an engine into that same space—I would probably build pretty much the same one again. Perhaps simplify valve operation a little and certainly use plain bearings . . . But it was a very healthy engine so I'd take the same path."

In truth, few engines have been so right almost literally from the start.

Jerry Sloninger

Engine Development

It is fitting—but also virtually inevitable—to explain the most successful race and rally cars of the postwar generation in terms of their engines. Bodies didn't change all that much but the one basic engine displayed near-infinite versatility.

For that matter powerplants always came first at Porsche, often before they bothered about the chassis. Their four-camshaft four was completely in step with company habit on this point. Whatever its passenger car antecedents the 547 would be very different in fact, which fed the mass of legend and rumor which grew up around it. By 1952, Ferry Porsche and team began to wonder if special breathing methods could achieve a then unheard of 70 hp/liter. Fuhrmann was told to find out.

The four-cam design he proposed then went from 70 to well over 110 hp/liter in a single racing decade—from certain class winner to overall threat and from pure racer to quirky road car which pushed out more sheer brute power in the Sixties than its Spyder forebear had managed 10 years earlier.

Porsche must have known Fuhrmann was the perfectionist who ignored means which hid his end. A master mechanic from the race shop recalls that the designer, "hardly ever let us go home before ten or eleven at night. He was an idealist. Never satisfied. And never afraid to pick up a wrench himself either."

By the fall of 1952, they were so far along a second decision had to be made. Should a small firm living in a borrowed corner of Reutter's body shop invest in such exquisite complication? For that matter, could they? In any case, Porsche did.

Their reward was an engine with four cams (two per side) atop hemispherical chambers, with dry sump lubrication, twin ignition, a pair of dual-throat carbs and a roller bearing crankshaft. This assembly reached the test bed on Maundy Thursday, 1953, a happy three years to the day after the first Stuttgart-built 356 of 1100 cc and a rousing 38 hp left their shop door.

Now they were looking at 112 hp/6400 rpm on the first try.

The built-up, roller-bearing crankshaft, built by Hirth, received considerable attention but it wasn't chosen for low friction only. In fact many felt the self-aligning unit increased drag due to fluctuating angular

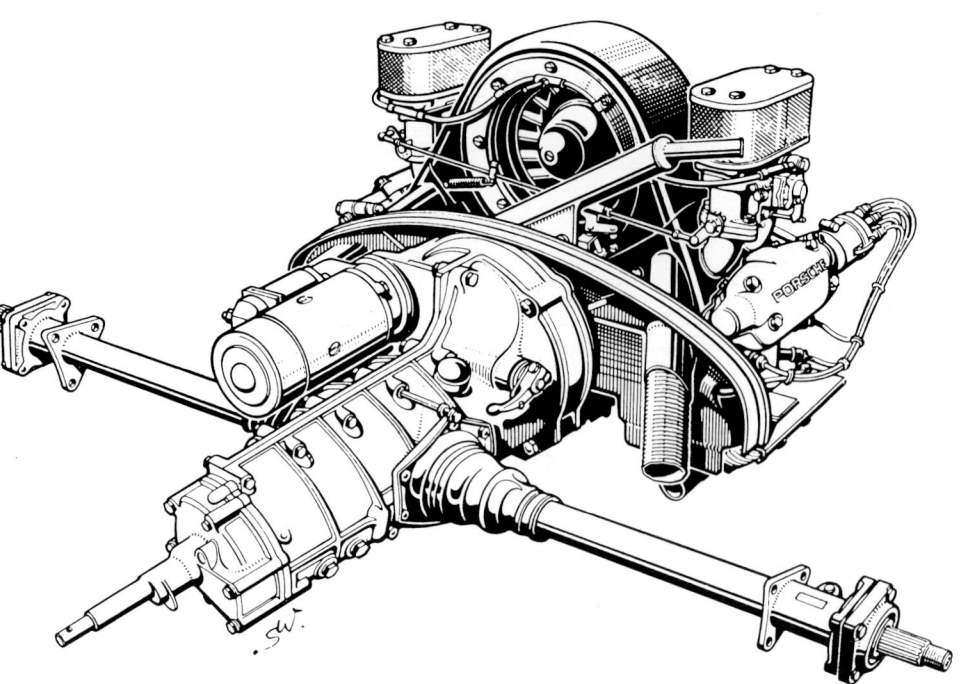

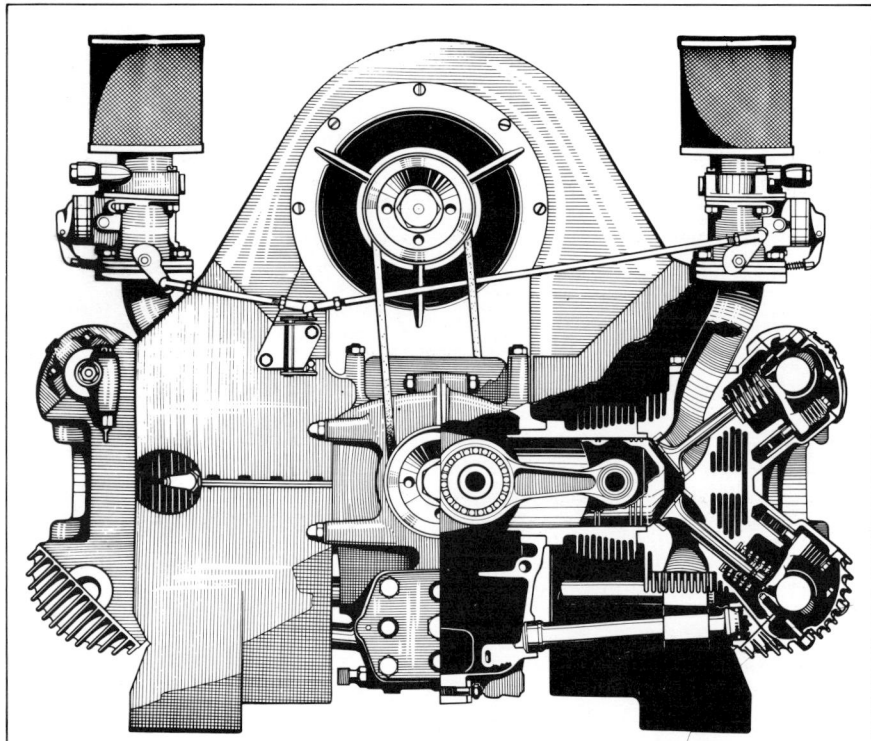

Layout of 550 engine shows crank, rod, piston and valve gear positions, below, and relationship to transmission and final drive at right. Opposite page: a Hirth crankshaft assembled and ready for installation in the racing shop. Weitmann photo.

accelerations although it did allow stronger one-piece connecting rods of lighter section. And it took up less space without sacrificing strength around the bearings.

As Fuhrmann said, "people thought in such terms then, which was a carry-over from the Thirties." Cranks like the Hirth were very robust for short periods but they lacked durability in the long run. Porsche's was designed with thicker crank webs and cylinder centers increased by 0.8", to 4.8", and crank failures still weren't unknown. They learned eventually that the suction pump wasn't large enough to pick up foamed oil.

Air cooling was a Porsche tradition but Fuhrmann chose a double-entry fan from the first for the only such engine raced seriously at the time. This absorbed some 8.8 hp/7300 rpm (versus 8.6 hp/5500 for a 1.5 liter 356 with 50% less cylinder area). The 547 used just 6.5% of its power for cooling and Porsche insisted conventional radiator systems would cost 20 hp.

This dual-entry fan was called the cleverest trick of all but it drew less attention than valve drive by nine shafts, 14 bevel gears and two spur wheels. This masterpiece of applied motions was so compact Fuhrmann never considered single overhead cams. He did run the kingshafts in plain bearings to save space, despite the added noise when the 547 type engine went into a road car. Noise dictated spiral bevel-cut rather than straight gears though.

From the crank, drive went via spur gears to a half-speed shaft (also turning the oil pump) and was divided by bevel gear sets and layshafts to turn the exhaust camshafts. Further shafts went up to the inlet cams which originally drove the twin distributors as well. As Fuhrmann said, "this was common then and with a 12 cylinder, shaft drive is very smooth. But I had only two cam lobes per shaft and their uneven motion ruined the coupling between cam and distributor very quickly. It was one of my mistakes." In addition, it also destroyed the ignition advance mechanism.

The designer still doubts that this cam drive system—properly adjusted by a master—absorbed any more power than a conventional gear drive train.

Intake valves were inclined at 39 degrees, exhausts at 40 degrees with dual springs. Valves had to be large with only two per cylinder, and this led to dual plugs to achieve decent combustion in that large bore. Asymmetrical lobe shapes compensated chiefly for finger followers and clearances were manipulated more easily than was customary for a race engine. The valve contacted its finger at the tip while the cam met it midway. Clearances were altered by moving this pivot.

At that time German race rules still allowed alcohol—Porsche tried it in one four-cam four—but they realized that international pump fuel requirements would mandate all this better breathing effort. Main

Spyder engine on a stand in the assembly shop (Weitmann photo) and a schematic drawing of the valves, camshafts and kingshafts.

1. Cooling fan housing
2. Vent pipe
3. Flywheel
4. Crankshaft main bearing
5. Crankshaft
6. Driveshaft for idler
7. Crankcase
8. Throttle linkage
9. Generator
10. Connecting rod
11. Distributor drive housing
12. Angle drive for steering with oil pump drive
13. Muffler
14. Oil pump

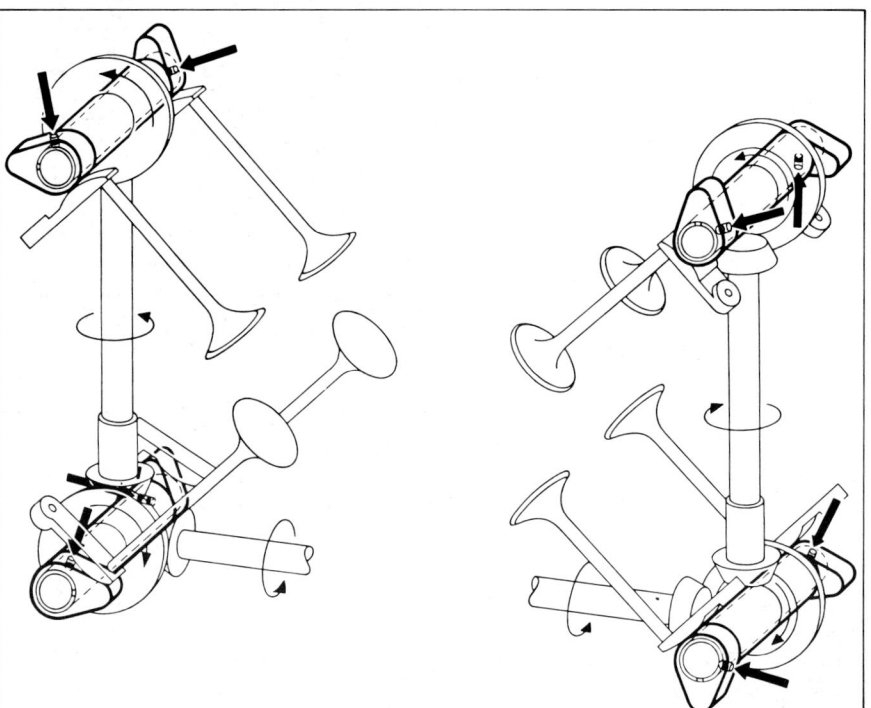

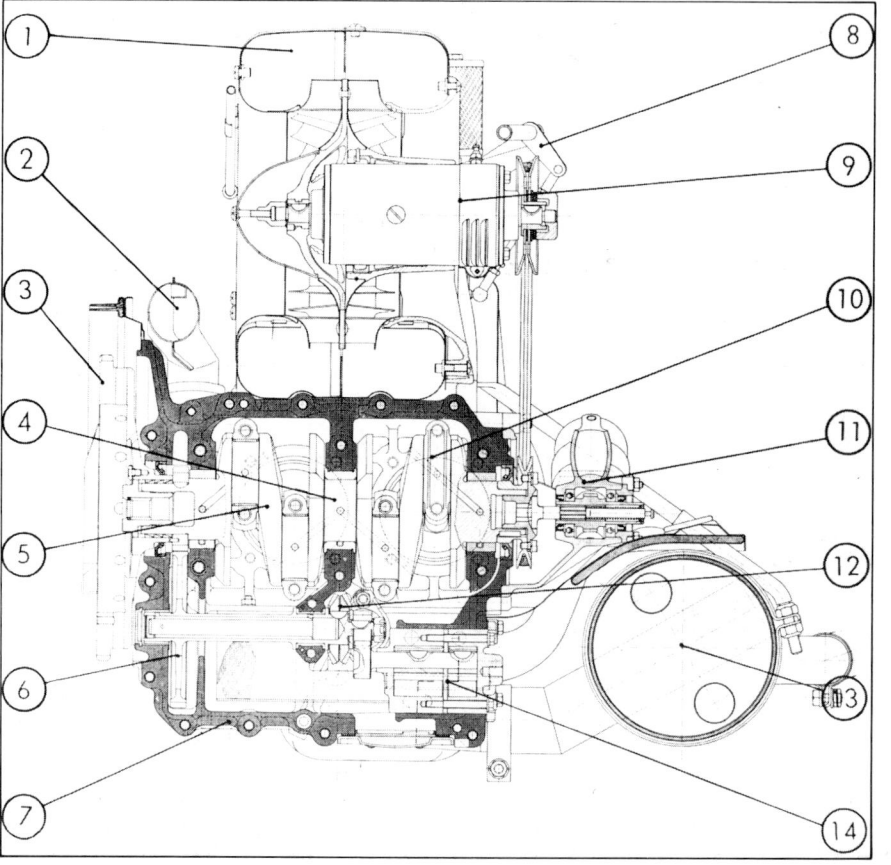

Top side (intake ports) of the four cylinder, four camshaft engine. Weitmann photo.

bearing and valve drive design here owed a fair amount to their own Auto Union and Cisitalia GP projects.

Weighing some 310 lbs. dry, this original 1.5 liter 547 engine would eventually go from 110 to 180 hp and was not particularly sensitive to mild overrevs although 7000-7500 was the initial limit.

Problems were remarkably few, given the complication, and haste of design. One of the first and last problems was oil forced out by blowby pressurizing the crankcase. Fuhrmann feels he chose the wrong pistons and rings.

It took a highly-skilled man about 120 hours to assemble each engine—the reason few saw any production car future for the 547. The meister recalls, "with good luck you might set the timing in eight hours. It could take fifteen too, if tolerances weren't just right." Those first race engine specialists later trained another 20-25 assemblers picked from Porsche's best when the road Carrera came along.

And when the roller-bearing crank was swapped for one with lead-bronze big-end plain shells and lead-impregnated aluminum main bearings the job became easier yet.

First, however, they assembled and tested the original maze and rushed it into a Spyder (which had to start its 1953 season with pushrod power).

The Porsche four-cam didn't win its first event (Freiburg hillclimb)—a pushrod Spyder and a Borgward beat it—but did hit the stride they expected by its second start and seldom faltered thereafter, although it was neither as light in weight as popular opinion believed nor blessed with any great power edge over the competition.

Relatively few survived in original form for the paradoxical reason that they were all so similar, barrels could be swapped easily to an older crankcase giving an old 1.5, say, instant 1.6 liters displacement.

550 Spyder

Porsche's first real racer, project number 550, eased onto the scene rather than bursting out in full four-cam cry. Early press comments lumped it with such 1953 peers as the Salmson 685 and Hudson Jet: as an open 356 prototype with two adjustable seats and an "all weather top."

It was too early to appreciate the car's chief virtue—balance. Neither as fast as the big-bore cars nor as weasel-like as the ultra-lightweights, this 550 was built to out-maneuver the first and out-last the rest. Since the first customer models would cost less than $7000, Porsche had also discovered value-for-money racing.

The firm likes to trace its genealogy to the prototype 356 with mid-engine but the more direct forebears were Walter Glockler's Frankfurt homebuilts. These alcohol-fueled, 980 lb. streamliners cleaned up both 1100 and 1500 classes and showed what a "Spyder type" car (the name came later) might do.

By 1953, Porsche management realized that Borgward, the Italians and East Germans were too fast in special racers for hopped-up 356s so they too turned to open racers. Their first 550 was even built at Weidenhausen in Frankfurt, the firm which shaped Glockler's cars.

Helm Glockler (a cousin) drove the first one at the Nurburg Ring Eifel race in May 1953 and beat the Borgwards in pouring rain with a 98 hp 1500 S pushrod engine running on alcohol. That was enough to send a team to Le Mans where they were limited to pump fuel and a mere 78-80 hp. Porsche turned to cramped, airless, nearly-blind tops on two cars which finished *ex aequo* first in the 1.5 class and 15th overall at 86.3 mph. The twins were timed down Mulsanne Straight at 122 and 124 mph.

Further wind tunnel work indicated that turning drivers claustrophobic had gained little. If the C_x value was down by 20% with a top the C_x times frontal area figure was only marginally better and they could achieve the same thing from an open car carrying full width windshield and headrest.

These first five 550s are often confused as Porsche juggled shapes. Chassis 550-1 was the open Eifel race winner given a top for Le Mans to match 550-2. This pushrod pair was sold to Guatemala and one took its Panamericana class in 1953 when the works cars retired. Car 550-3 carried the first 547 four-cam engine during Nurburg Ring practice in August 1953. It didn't race but cleverer fans noted a different sound.

One week later this car made its race debut at the Freiburg hill-

The original Glockler Spyder (Nurburg Ring, 1950), winner of the German title, and forerunner of the Porsche 550. Weitmann photo.

climb, Stuck driving, but managed only third when he had some trouble with the power band. Fuhrmann comments, "we were all a little disappointed. With 30 more hp than the pushrod engine it didn't go all that much faster. The real reason was simply that you can't put a new engine into an 'old' chassis."

Back in pushrod form 550-3 crashed out of the Liege-Rome-Liege marathon.

Car 550-4 was the "humpback" version, a 547 test bed also shown at the 1954 Brussels Motor Show. Chassis 550-5 had the shape eventually chosen—no hump, one-man windshield and fairing—and it was shown at Paris in the fall of 1953, before the 550-4 in fact. Weidenhausen built the first works prototypes, Weinsberg the next variants.

All had ladder frames but the first pair carried their transverse torsion bars behind the rear axle line while 550-3 and thereafter ran the tube ahead of the axles like their road cars although the racers used a longer 23.6" trailing arm. Rear shocks were more perpendicular and half-shafts ran below the kicked-up rear frame, unlike 550-1 and 550-2.

The 1954 season opened well when Herrmann in a 550 not only won his 1.5 class in the Mille Miglia (aided by broken Oscas) but was an amazing 6th overall despite 20 minutes spent drying the electrics. That race held the famous corner which he rounded to find the hidden railroad gates were down. Smacking Linge on the helmet Hans ducked too and proved how low a 550 was by going under the bar.

They may have been too cocky for Le Mans. All three works 550s burned pistons due to faulty ignition settings. Yet one limped home on three cylinders to win the class—luck as well as durability rode with Porsche.

Almost forgotten in that fiasco was number 47 which took a clear 1100 class victory. It also had a four-cam engine, but Fuhrmann recalls, "that never produced the power we expected. We thought big-port 1.5 heads would be fine, for maximum power at least. Now I know we should have used smaller valves. The class wasn't so interesting to us anyway."

Two weeks later 1.5 Porsches got so far in front of the Reims 12 hour race they whipped 2 liter cars while running under an "EZ" signal. Spyders formed a 4-car front grid row on Nurburg Ring and finished the same way. Then Porsche gave Herrmann another Panamericana shot and he took his class (3rd overall) from a sister 550.

To beat a Porsche in class you needed another 550 and the US importer was crying for just that in 1954. They had always planned a small series and named them Spyders in honor of open American sporty cars of the Twenties.

These "production" cars were built by Wendler of Reutlingen with

Glockler Spyder brought into U.S. by Max Hoffman, later ran at Torrey Pines, California in 1952, driven by Ed Trego. Cockpit is shown on opposite page. DB/Poole photos.

Jaroslav Juhan's 550 Spyder finished 2nd in class, behind Herrmann's similar car, and 4th overall in 1954 Panamericana. Coltrin photo.

Johnny von Neumann at Pebble Beach in 1954, in one of the first 550 Spyders seen in the U.S. Poole photo.

Fernando Segura, of Argentina, drove this pushrod 550 coupe to 2nd in class and 33rd overall in 1953 Carrera. DB photo.

nose panels tucked under and a slight fin look to the rear fenders. Following 15 chassis numbers reserved for works cars the next 75 were assigned to customers, mostly Americans. They were very like the factory machines with the same 110-112 hp engines but a 1300 lb. weight against 1215 for the original cars.

Built in the same race shop they used the same, new side-rail stiffening found necessary for decent handling from what amounted to VW suspension: twin front trailing arms and swing axles in back.

Customer engines, designated 547/1, had a hair more torque although all such figures depended on which spec sheet you read. Compression should have been 9.5:1 for all but as low as 8.7 has been quoted and power was purposely underplayed to give buyers a happy surprise should they ever tame all those bevel gears.

Even as Spyders went on sale, one English weekly still called the 550 a "modified 1300" but despite their faint production aura these were certainly one-off racers with no two precisely alike. For instance, many believed all Spyders had the center-lock wheels shown in Paris although Porsche found the car was too light to need tire changes in racing and only made the one set.

With production moving, Porsche opened 1955 with six class records at Monza, taking marks from 200 miles to six hours with a Spyder which lapped at 130 mph.

They finally got Osca's measure at Sebring too when six of seven Spyders finished and that orphan 1100 four-cam won its class as well. It would repeat at Le Mans in 1955 while an older '53 Spyder 1500 headed a 1-2-3 class sweep for Porsche with first on Index as well to avenge their 1954 debacle. On Mulsanne Straight the 1.1 liter car did 118 mph and the open four-cam 1.5 clocked 139.4.

During 1955, class wins were almost commonplace: in the 1000 km of Paris, the 1000 miles of Brescia or in Venezuela where they flew one car in, drove out to the track and harassed the big boys all day.

Back home the portents were less rosy. An East German EMW trained faster on the high-speed Avus (despite 125 hp in the Porsche) and a Spyder won only when the best EMW fell out. At Nurburg Ring a 550 was straining for 10:50 laps while Barth's EMW and Behra's Maserati were lapping at 10:30.

Clearly Stuttgart's "alu-foil specials" had lost their easy edge. Porsche even let visitors drive them on the nearby autobahn and offered rally fittings for the cockpit. Spyder novices found they could pull away from as low as 2500 rpm yet do 0-100 in less than 20 seconds. The pros wanted more.

In California, Ginther was getting 130 hp from 12:1 compression

The "Humpback" with center-lock wheels. This was 550/4, the first seen at a motor show, but not followed up.

Ken Miles in von Neumann's 550 after Ken quit driving his MG specials. Poole photo.

Karl Kling in the 1953 Carrera Panamericana, driving one of the early pushrod Spyders. DB photo.

Hans Herrmann's pushrod Spyder went out on the second leg of the 1953 Carrera with a broken front spindle. DB photo.

Spyder with enclosed cockpit, one-man windshield, allowed by rules in early Fifties. Weitmann photo.

and Weber carbs—curiously, just about what works engineers had achieved with a 547 on alcohol. With one-man windshield this car did a true 128 mph and 18.5 mpg in a "road test" which included "hard driving." Yet it was driven to the track from downtown Beverly Hills carrying wiper, turn signals and speedometer.

The Spyder, once called fiercely powerful, was now rated "fairly fast but easy to race," although Ken Miles felt that only a genius would extract the ultimate performance. The transition between tail end going and gone was very small since many overlooked the swing axles until hitting a bump in the middle of a fast bend. Rear unsprung weight met front suspension flex to make a handful.

The smoothest, straightest possible line won in a Spyder which was noted for light steering and pedals, a soft ride and the fact that all noise and heat was behind you. This prompted Miles to call 550 the finest long-distance racer in the world, regardless of capacity. For long events most used an "overdrive" 0.88 IV gear (any ratio could be changed without disturbing the rest) and moved III up to normal direct, making low a starting gear only.

That minority of US racers who didn't overrev—most shimmed the valve springs and went 1000-1500 over factory limits—found the 547 powerplant relatively bulletproof too. Everything was adjustable, thus it was easy to get one out of adjustment. But it was equally easy to work on, "provided you took care to put it back together as original."

Nonetheless, 550 racers were beginning to face English flyweights and US events were too short generally to exploit Porsche virtues. Stuttgart sought the answers.

Studiously ignoring the extra speed which Swiss privateer Mike May gained for his Spyder by mounting the first-ever wing on struts to beat even works cars during Nurburg Ring practice—he wasn't allowed to race—Porsche engineers went for increased power instead. A few more revs, a whole new frame and a five-speed (really 1 plus 4) gearbox gave them a new car.

This was the Spyder called either 550A or 1500 RS with their usual disdain for consistent nomenclature.

550A/1500 RS

Porsche opened 1956 with a two-pronged program. Specification sheets for their customers continued to tout the 550—after all if the charming Gilberte Thirion in a Spyder could beat a 300 SL and a Ferrari 250 GT, both driven by GP class men in a hillclimb, the car couldn't be hopelessly outmoded.

Meanwhile the works could devote full attention to a more specialized new Spyder with an admitted 130 hp to equal their best engines from the end of the previous seasons. Herrmann/Trips used such an engine in the 550 chassis for a class win, 6th overall and 1st on Index at Sebring, covering 181 laps or only one less than the winning D-Jag a year before.

Over the winter they merely mated this with a chassis so different it really deserved more than the suffix A.

With some justification, the Porsche operation has been described as a good hotrod shop guarded more closely than any Detroit styling studio. By 1956 they were building both the 550 which now ran as a production sports car and the 550A which would be their "race" model.

Apart from extra power in the A, the chief difference was a proper tubular space frame replacing the ladder and cutting weight by 35-95 lbs. It also allowed them to pare another 60 lbs. off the body weight (down to 140) by providing more panel attachment points. At 1170 lbs. the cars were thrice as resistant to torsional forces and five times as strong in the bending mode.

Sole front suspension change was a stabilizer but the trailing arms worked better in a stiff chassis while steering became Porsche's worm and sector. In back the chief change was low-pivot swing arms with the two halves jointed below the differential in Mercedes manner.

Along with larger tank (34.3 gal. replacing 25) this 550A had wider front brake drums with peripheral ribs—as introduced on Le Mans Spyders in 1957 where they estimated that 1700 hard applications in 24 hours had produced 1mm of front shoe wear.

Finally there was the almost-five-speed gearbox. This was their familiar four-speed plus a start-line low in a nose casing, without synchro. This allowed them to space the other four closer together. Porsche offered three final drive ratios and three gear sets for each of four ratios and once their closer-coupled shift pattern was sorted out the overall gain was considerable.

Customer 1500RS with full-width windshield and wipers, was the second series Spyder with bodywork similar to original. Werkfoto

Left, and above, John Porter's "customer" 550/1500RS didn't have the "A" suffix as ladder-type frame was still used while factory cars had new space frame. Poole photos.

Ricardo Rodriguez in his 550A/1500RS at Riverside Raceway. Poole photo.

A 1500RS races a Klemm in the traditional car vs plane contest made popular by Barney Oldfield and others. Weitmann photo.

The 7:31 ring and pinion set, giving a 4.428 final drive ratio, was "standard" for every four-cylinder Porsche.

The 135 hp engine (130 for longer events, indicating they suspected its limits) had 9.8:1 compression ratio now and a smoother torque curve with higher overall lb-ft rating but no basic features were changed.

Wendler of Reutlingen made the bodies as they had for most 550s and the A series ran to seven works cars plus 30 private Spyders during 1956/57 although the factory eventually held back one of the customer chassis to make eight. Dimensions matched the 550 with one car excepted.

Nurburg Ring lap times best show the chassis' worth. Despite notable power increases since 1953 lap times hadn't fallen much in the old chassis. Now they saw 10:30, an improvement out of proportion to the added power. Drivers had an easier time too since tail break-away was more progressive.

The 550A bowed in at the Nurburg Ring 1000 km race alongside "production" 550s and Carreras which won their respective classes. In the race division von Trips/Maglioli and Herrmann/von Frankenberg beat the East German AWE (ex EMW) as hoped and this encouraged Porsche to send one car into an enemy camp where it recorded the finest triumph to that time.

Maglioli would try the Targa Florio, driving solo and supported by a race boss and two mechanics. With just one pit stop and no tire changes the jut-jawed Italian grabbed the lead on lap two and held it to win the second-fastest Targa up to then at 56.5 mph. And he did it on just 1500cc.

A euphoric Porsche immediately turned to Le Mans where higher windshield rules made an enclosed car logical. One such 550A "Coupe" brought their sixth class win in a row, 5th overall and 2nd on Index but the second car retired with engine trouble. The class winner covered 2356 miles, much of it in the rain, yet returned 17 mpg.

Spyder construction at Porsche. The 550A/1500RS was the first Porsche to have space-frame construction, which lowered the car weight (dry) to 1170 lbs. Weitmann photo.

A 1-2-3 in the 12 hours of Reims, again beating the 2 liter Ferraris, added to a banner year which was supposed to culminate on Berlin's dangerous Avus Ring. Porsche overreached a little there. The background was laid on Solitude, a track on the factory doorstep, where works Spyders finished 1-2 from an AWE while their "test" Spyder came 4th (and production 550s were 1-2-3-4 in their class, incidentally).

This test machine was a further step beyond 550A, a car now ready for production itself. Wheelbase had been cut for the first time, to just 78.8" which happened to match the Glockler car and was nearly 4" shorter than a 550. Track was reduced by several inches as well and the slimmer car nicknamed Mickey Mouse by works mechanics.

Frankenberg drove it at Solitude and noted that, "handling was not the best and the brakes gave trouble." It trained but didn't race on Nurburg Ring and then went to Berlin where the smaller body, faired headlights, covered second seat and extended nose and tail were expected to give them an edge.

As von Frankenberg put it—"we'll have to see what these new dimensions do to high-speed handling."

In a race won by von Trips' normal Spyder Frankenberg discovered all too clearly how a shorty might handle on a bank with a kink in it. He made one of the most spectacular exits ever photographed, flipping over the top of the high bank. The magnesium-bodied car crashed in glaring white flames but the driver was pitched clear to survive with light injuries.

Thus Porsche entered 1957 with the old, familiar dimensions, dropping M. Mouse from contention (and all conversation) although its oil cooler surfaced again in the 550A's successor. Instead they devoted the winter following their final '56 event to building 30 A replicas for eager customers. The factory would run an A for part of 1957 as well, until the new car was ready.

Barth at Freiburg-Schauinsland Hillclimb, July 27, 1958. Coltrin photo.

Count de Beaufort at the 1958 Nurburg Ring 1000km race. Molter photo.

Weitmann photo

Mickey Mouse was a very special Porsche; handbuilt, with shorter wheelbase, narrower track, and tricky to drive. It almost cost von Frankenburg his life at Avus, and after the crash was never heard from (or talked about much) again. Weitmann photos.

Race speeds were soaring now—at Reims Carreras lapped faster than Spyders three years earlier. But private 550As were still fast enough to win at Le Mans for Porsche when works cars crashed or ran out of fuel. Then a Lotus-Climax win at Spa with an Osca third proved that purpose-built lightweights could beat a 550A, even on the continent, so work was hurried on the RSK, as seen at the Le Mans race in 1957.

Their racing spectrum broadened during 1957 too. Edgar Barth, who had moved west when AWE dropped racing, was entered for the F2 portion of the German GP in a perfectly normal Spyder and actually beat Salvadori's open-wheel Cooper. It set Stuttgart thinking hard about formula racing.

Meanwhile the once-famous European Hillclimb Championship had been revived in 1957 after considerable German urging, although events were held for two liter two seaters to suit Italian wishes. Porsche sent a 1500 RS anyway and soon were using these alpine sprints as basic test grounds for all their new Spyders.

The very first Porsche move out of "their" 1500 class came when von Frankenberg was sent to the Austrian hillclimb with a 1588cc engine in his Spyder. By the end of August von Trips was running 1680cc, then considered about the crankshaft limit for a 547 four. Germany's coming star was 2nd to a 2-liter Maserati in his first try, then won in Greece with this 170 hp 1700 RS.

However Porsche had gotten Trips started late in the season and rules specified a driver must contest four events so a Maserati and a Borgward were 1-2 in the 1957 uphill chase and Porsche wrote the season off to experience. They came back strong in 1958 of course and by then were using the new Spyder which had really been a mid-year '57 model, the RSK.

Victor Rolff's 1500RS at Hockenheim, Germany, in 1956. The bubble top lost as much in wind resistance as it gained in aerodynamics. Weitmann photo. Jack McAfee, in John Edgar's 1500RS, with modified bodywork, at Riverside. Poole photo.

718 RSK

The casual race fan who was likely to lump the 550 and 550A as those 1500 look-alikes soon recognized the RSK as that new 1600 with fins. As a generality it was reasonably accurate.

Porsche introduced the new model into the middle of the 1957 season as a works prototype which was soon known generally as RSK, internally as 718 and occasionally as 1500 RSK 1959. Consistent in their endemic inconsistency they always called it RSK—in honor of revised front torsion bar carriers in the form of a supine K, although this suspension was abandoned well before '57 ended.

A new project number was curious in other ways too. Whereas an all-new space frame for the 550 only rated an extra A, adding/deleting a tube from that same space frame merited the model number 718. Admittedly most of the 1959 changes occurred beneath a familiar skin.

Outwardly the RSK was some five inches lower and sufficiently sleeker to have 10% less wind resistance although never quite as tautly wrapped as Mickey Mouse. They had returned firmly to the known wheelbase and track—and finally settled on a change to the rear suspension to honor a car named for a front suspension it didn't use.

On its first appearance, at the German 1000 km race in 1957, the car featured turbo-finned brakes in front, ribbed drums in back, wheels with larger cooling slits, the flat-K torsion bar carriers and ball-jointed front trailing arms spaced much further from one-another. There was no oil cooler slit in the nose since the 718 used a cadmium-plated surface cooler in the front lid, a nightmare to solder up, but efficient. Their best car in that race was an old RS, 4th overall and 1st in class.

For Le Mans the works RSK suddenly sprouted fixed fins on each rear fender which didn't seem able to cope with an inherently low polar moment of inertia. No two drivers agreed on their worth—where one wouldn't even sit in a finned car, another wouldn't leave first gear without them. But a 550A did their winning.

By the end of 1957 Porsche was ambivalent about fins and more intrigued with a possible replacement for their double-entry fan. While that only consumed 8 hp this was still 5% of available power and Porsche admitted to experiments with "jet cooling," a forced draft of exhaust gas used to suck fresh air over the cylinders. It all began with a Fletcher project in America for a Porsche-powered paratroop vehicle.

Its fanless cooling conversion caught the competition department's

The RSK, or 718, rolled out as a works car in mid-1957, and carried Porsche's colors through 1959. Werkfoto.

First race appearance of the RSK was at the Nurburg Ring 1000km race in 1957. This is Maglioli during practice. Molter photo.

Graf von Trips in the RSK at Nurburg Ring's "Sudkehre" in 1959. Molter photo.

RSK cockpit, 1958, with foam padding for driver's left leg. Weitmann photo.

Porsche garage at Le Mans, 1957. Fins, seen on the RSK for the first time, met with mixed driver reaction. Coltrin photo.

The Frere/Schell finned RSK at the Nurburg Ring, 1958. Coltrin photo.

eye but the announcement had proved premature and nothing more was heard, apart from rumors which dogged them clear through 1958.

The '58 season opened with a 1000 km race in Argentina only a week after Moss gained the first post-war GP victory for a rear-engined car. He was scheduled to drive a Maserati in the two-seater event but when that expired in practice Moss became Porsche's second pro. Behra was already hired for the season, a move which finally put Porsche into the paid racer category.

With a little larger engines they could come close enough to overall glory to attract GP-caliber talent. Also, the 718 was sensitive to handle and needed such skills to extract its ultimate.

In fact Moss/Behra didn't drive a K although they had the 1.6 liter engine and finished 3rd, only 8 seconds behind the leading Ferrari, winning the 2 liter class while a semi-private 1.5 won its class in 5th overall. By using the ex-hillclimb engine, Porsche ran in two classes through much of the season.

At Sebring one RSK 1600 lost a rear axle oil seal after putting 100 laps into 6 hours (the eventual winners did 200 laps in 12). Schell/Seidel completed 193 laps in Porsche's second car for another overall 3rd to cars of twice their capacity.

The real news by now was coil springs replacing torsion bars for the first time—a case of the race department outvoting their sales side. Coils eliminated the heavy torsion bar carrier. Low-pivot swing arms were located by Watt links at their outer ends while pot joints replaced sliding splines on the axle shafts.

A 35% weight saving was also the reason Porsche gave for drum brakes. They felt no disc showed an advantage over their own alloy drums with cast iron liners rubbed by Energit shoes of steel wool and bunyl rubber. Since they had to remove front drums at both Sebring and Le Mans for pit-row brake repairs, however, discs could not be far behind.

Engines were largely unchanged for the year apart from the 87.5mm bore option which gave 148 claimed horsepower with flash readings up to 160 for this 1.6 while the 1.5 now had 142 hp. That 9.8:1 compression

Start of the Nurburg Ring 1000 km race in 1958. Moss does his usual fine start in Aston Martin No. 1, while assorted Porsches, Aston Martins, Ferraris and a Lola follow. Molter photo.

Finned RSKs were also driven by Barth. Weitmann photo.

ratio led to large valve clearance cutouts in the pistons but fuel injection was no longer mentioned publicly. And the surface oil cooler was supplemented by a second one on the cockpit floor with its own air scoop.

Behra achieved the next success with this lively, finned RSK 1600, finishing 2nd overall to works Ferraris in the Targa. (The 1.5 class was left to privateer shoals wherever possible.) Porsche insisted they weren't seeking overall wins and were "just lucky." They weren't even that before the home crowd which now expected no less than a 3rd overall if not a victory. The works seemed to push too hard at times, trying to oblige.

For the Nurburg Ring race they even ran team 1500s to beat back dreaded Borgward and won as planned with an honorable 6th and 7th overall. Behra even lapped at an amazing 9:54, with fins. For Le Mans the 1.5 had fins (the 1.6 did not,) but Porsche reduced compression to 9.5:1 and garnered a 3rd to a Ferrari and an Aston Martin with their 1600 while a pair of 1.5 Spyders in line astern took that class as well.

Behra obviously preferred fins, using them on a center-seat RSK which won the Reims F2 event from a Ferrari open-wheeler. Ferry Porsche had said they would run a limited number of the 1.5 liter formula

Bonnier hangs the tail out at Riverside's turn one, October, 1960. Tronolone photo.

Wood body buck in Wendler's shop in Reutlingen. Weitmann photo.

Freiburg-Schauinsland Hillclimb, July 27, 1958. Behra's RSK finished 2nd overall. Coltrin photo.

races "on fast circuits." The basic RSK had been designed with center-point ZF/Ross steering which made a single-seat version easy. For these shorter events the Spyders produced 164 hp.

The hillclimb series was contested by 1.5 cars that year too. Three Spyders beat two Borgwards in France only to find Behra second to a Borgward at the German meet while von Trips beat their north German rivals in Italy and Austria, clinching the 1958 uphill title.

Behra now decided Trips was getting better cars from Porsche although he had lowered his own 1.5 sports car mark on Nurburg Ring to 9:48.9 that fall. Still, Jeanot was convinced he was being done despite his star status so they swapped cars for a Zeltweg airport race and staged a barn-burner which Trips won by 0.8 seconds. Behra settled for the title of "1958 German Sports Car Champion."

To close their home season, Porsche sent a full team to Avus—notably without fins despite the flat-out nature of that track. In their best faceoff against Borgward all year Behra won by 0.8 seconds although Bonnier's front-engined Borgward went visibly faster and appeared to handle better.

Dutchman, Carel de Beaufort was getting the hang of his own Spyder now and won the 2 liter class from a Maserati. Then they discovered his powerplant only displaced 1.5 liters and disqualified the genial Dutch giant for "too little engine."

Porsche sent one RSK 1500 to the English TT and took 4th, leaving them tied with Aston Martin for second among 3 liter race cars in 1958. Porsche scored points in all six events but only the best four finishes counted: theirs were a second and three thirds overall.

Stuttgart also registered its baker's dozen of national titles around the world plus a couple of Austrian records—and even won a revival of that old fairgrounds race between car and airplane by a nose over the Klemm.

More important was an FIA decision that Grand Prix racing would be a 1.5 liter affair for 1961—this despite howls from the English who had been trailing converted Spyders in 1500 F2 events. For 1959, Porsche would build a proper open-wheeler, launching Porsche's monoposto era which almost certainly slowed Spyder development.

The two-seater weapon for 1959 was still called RSK. Now producing its power at 8000 rpm with an 8400 limit (9000 very briefly) this car

Opposite page. Ken Miles at Riverside in the Estes-Zipper RSK, October, 1960. Tronolone photo. Rear suspension of RSK. Molter photo.

At Le Mans in 1959, the works cars were supplemented by US. "privateers." Sloniger photo.

Chuck Howard at Laguna Seca in June, 1959 (75) and Bob Holbert at Riverside in July, 1959. Tronolone photos.

would do a true 160 mph or break 9 seconds for 0-60, depending on gears. This brought such an RSK within a whisker of a class standing-start kilometer record held by a supercharged 1.5 Maserati although the spyder clutch would only take four such drag starts while testing the idea.

Porsche quoted 148 hp for private cars but few left the plant showing less than 150 on the house dyno. By the spring of 1959, they had turned out another 22 RSKs for private owners, adding to 135 Spyders already built. Only five remained in Europe, one a center-seater, and the remaining 17 were sold to Americans for some $8000 each—explaining unfilled orders for another 80. Their shop barely found space to build four at a time as it was.

Drivers who did get one found rather heavy pedal pressures but very light, quick gearbox (still 1+4) and a continuing tendency to terminal oversteer despite progressive coil springs. Works pilots insisted handling was so close to neutral they were getting understeer in downhill bends. In short: somewhat more difficult (again) than its predecessor.

Yet you could screw a baffle can "muffler" into the tail pipe and drive your RSK around town at 1800-2500 rpm on the warmup plugs, no

Le Mans, 1958. Porsche finished 3rd (Behra/Herrmann—29), 4th (Barth/Frere—31) and 5th (de Beaufort/Linge). Goddard photo.

Goodwood, 1958. Behra in an RSK with no headrest as usually seen on the type. Ross photo.

Behra again, at Riverside in October, 1958. After qualifying 13th, he finished 4th in a smooth drive overshadowed by Daigh's win in the Scarab-Chevy V-8. Tronolone photo.

small feat with nearly 100 hp/liter from a six-year-old racing engine. Brakes remained finned 11" drums since the works claimed 1g decelerations with them.

The 1959 competition record was mixed. Obviously second-string on occasion to the F2 toy, RSKs enjoyed tremendous high points like the Targa, fast becoming a Porsche preserve, but catastrophic troughs such as Le Mans, the most visible race of them all from a PR standpoint.

First Sicily: Porsche was fresh from a 3-4-5 at Sebring to Ferraris even though one car was trying new dual-wishbone rear suspension. This tended to lift its inside wheel in bends but still cornered faster than the low-pivot, swing-axle model.

The goal was reduced toe-in for the outside rear wheel. The solution was a short top wishbone nearly parallel to a bottom v-member with tubular trailing arm. Both had angled pivot points. The effect was that of angled swing arms but with a far-longer swing radius (nearly 10 feet). All arms had threaded adjustment at the frame for fine tuning.

Despite a mixed Spa detour where Porsche loaned an RSK to Frere but only de Beaufort's private car finished—in front of its class at least— Porsche journeyed to Sicily in good spirits.

Once there silver cars, led by Barth/Seidel the first German drivers to win since 1924, cleaned up although the only RSK at the finish was the leader, followed by three private and/or GT Porsches. Their 1.6 dominated from half-distance and set a lap record only to retire with gearbox failure just 25 miles from home. Overall victory thus went to their 1.5 class winner.

If we overlook obvious signs of engine strain, Porsche did well at Nurburg Ring too with 1-2-3 in the two liter class and 1-2 in the 1500 plus 4th overall but again it was a case of ample forces covering for failures. Most of the 1.6 engines expired during practice or pre-race warmup and top crews ran the 1500.

If the Ring was luck amid diversity, Le Mans was pure chaos. Ignoring all those 1600 engine failures Porsche picked a hot cam and not just for one "rabbit" but for all their cars. Engines failed right and left—all but Barth's and he had no clutch to get the power down anyway.

Things boded better on Avus with a sports car race tailored to their best 1500 capacity alongside the German GP. Porsche did claim 1-2-3-4 but Behra was killed when his Spyder went over the bank and only the most fantastic racing luck saved de Beaufort from a like fate. His Dutch

Ricardo Rodriguez' RSK at Riverside, July, 1959 (No. 3), where he finished 5th. Tronolone photo. Left, below, a customer RSK at the Nurburg Ring in 1959. Sloniger photo. A finned and auxiliary-lighted RSK for Sebring in 1958. Werkfotos.

Spyder jumped the bank too but landed on its wheels so Carel drove through a fence, the paddock and back onto the track to continue racing until droop-jawed officials black-flagged him for an inspection. Spyders were always sturdy.

Trips/Bonnier closed out 1959 with a second to the winning Aston in the TT race.

On the hills Barth swept all before him, often heading 1-2-3-4-5 Spyder finishes and claiming the title in a 1500 RSK. That fall they used a 1.6 liter engine of 152 hp to claim six 2 liter class records from 1 hour to 1000 km at Monza, giving the four-cam 547 nine such marks in all.

The habitual bag of national titles world-wide testified to the truism—if you want to win, buy a Spyder. Stuttgart did its best to continue this tradition in 1960, whenever their burgeoning formula ambitions allowed. You could make a good case that this RSK was the key model of all Spyders between 1953 and the early Sixties.

It brought a peak in Spyder suspension design by using wishbones and coils despite all house tradition. It featured their best 547 versions in all sizes except 2 liters still to come—and even that size would appear in an RS61 which really offered no further chassis advances of note. And this RSK displayed the full range of their streamlining thoughts in both two-seater and quasi-formula form giving them the frame, engine and impetus to enter GP racing.

One model could hardly do more.

When tire tests were necessary, Porsche took whatever car was handy, including this autographed car from the Frankfurt show. Sloniger photo. Behra (29) at Le Mans, 1958. Goddard photo. Works prototype at Nurburg Ring in 1958. Sloniger photo.

RS 60

By the time Porsche reached model designation RS60 in its Spyder hierarchy they were largely concerned with refining a proven design—further innovations need not apply. Unless you consider yearly model number changes as a new feature, which they were of course for Stuttgart. Also: private Spyder "replicas" now appeared in the same year as the works racers which spawned them, underlining the stepped-up pace of Sixties motor racing.

This RS60 was a sophisticated, efficient racer with its own features. For one it brought their first wheelbase change (ignoring the Mouse, as Porsche did). In this case they went 4" longer to give more cockpit space and more predictable handling. The rear frame unit unbolted for quicker ratio changes and better engine access. Wheel size became 15 inches.

Spyder bodies were still built by Wendler while each engine was assembled by one man, as they had been from the beginning.

Their plan called for a dozen private RS60s—two for Germany, the rest exported, largely to America which captured seven while two went to Canada and one to South Africa. It appears however that Porsche kept four and built 14 in all if we can believe chassis numbers. Thus they either built two not recorded or sold only ten. They never really made money on Spyder sales anyway and only sold enough to still the clamor.

The most notable external change was a taller, framed windshield of FIA-mandated height. A luggage compartment was also required—at first they tried to fit this imaginary case into the nose, later mounted it above the gearbox.

Engine figures had stabilized at 150/7800 for the 1.5 and 160/7800 for the 1.6 which was now their customer car since the class limit had been raised to 1600cc. These were numbers 547/3 and /4 respectively but there was also a works 547/5 of 1678cc and 170 hp. Having just missed the Manufacturer's title in 1959 Porsche tried again, using 1.7 liter cars in many events. The engine had survived its Alpine test and could always snaffle a 2 liter class win if overall laurels eluded it.

This system gave them a 3rd overall in Buenos Aires plus a 5th for a private 1.5 liter Spyder. Privateers had to use the 10" windshield too despite Porsche protests that this made things too hard for "several hundred owners." The very fact that there were so many potential Porsche winners about (around 200 to be honest) might have been one reason the change was jammed through FIA councils.

Barth/G. Hill RS60 finished 5th in the 1960 Targa Florio. Coltrin photo.

Introduction day for the RS60 at the Stuttgart works. As tidy as any new car before the battle, Molter photos.

At first the taller glass was reckoned to cost 5 seconds a lap or 6 mph in top—the factory hired GP drivers Graham Hill and Bonnier to offset this. At Sebring where Porsche, like Ferrari, abstained due to a fuel sponsorship hassle Bonnier then appeared as "entrant" with a pair of RS60s looking curiously like team machines and tended by mechanics speaking very Swabian German.

The Hill/Bonnier "private" car broke its crankshaft halfway through the 12 hour event but the second car, using only 7000 rpm of the permitted enduro 8000, came in first followed by a US Spyder. Thus Porsche went to the Targa Florio with the same Makers' points as Ferrari despite their displacement handicap.

Porsche left Sicily in front of the red machines from Maranello thanks to a second overall Spyder victory in a row. Sending a pair of 1.7 RS60s and one 1600, plus two Carreras, Porsche captured 1-2-5 and the 1600 class with the Spyders only just beating their Carrera mates. Small wonder Nurburg Ring drew a quarter-million loyal spectators, the largest crowd at that 14-mile course since the days of the pre-war silver steamrollers from Mercedes-Benz and Auto Union (the latter also a Porsche design).

Porsche didn't win for the home fans but did finish 2-4-6-7, led by a 1700 Spyder and spiced by the appearance of their prototype annular disc brakes on an Abarth Carrera. The best Spyder even led overall late in the race but an extra pit stop plus Moss magic in a Maserati was too much for the smaller-engined home team. Since the best Ferrari was only 3rd, however, Porsche stretched its marque points lead slightly.

Le Mans remained *the* race and Porsche was determined to improve on 1959. They entered a pair of 1320 lb. "1700" Spyders (actually a

The year of the high windshield was 1960. This is Gendebien at the Targa Florio. Coltrin photo.

Le Mans was not Porsche's race in 1960. Opposite, an RS60 goes through the esses and, below, mechanics performed a complete engine rebuild in the pits. Sloniger photos. Right, the RS60 engine compartment. Worner photo.

one-time effort with 1605cc from 88mm bore), three of the regular 1.6 RS60 of which two were open and one had a coupe top for more comfort but weighed 1765 lbs., and one 1500 seeking an Index of Performance win since it had to run in the same 1600 class as its three teammates.

Further detail work had reduced the penalty of that FIA windshield from 500 revs lost to only 100 for their "topless coupe" which also had a much needed extra inside windshield wiper with its own motor.

In the race this Porsche armada fell out one by one, the first 1.7 pitting after three laps with a strange noise in the left rear and the other breaking a rod. Porsche had added a fourth main bearing on the crankshaft nose and further limited revs to 7400 (their 1958 cars could rev to 8000 for 24 hours) but luck rode elsewhere and not even a complete engine rebuild on the pit cement helped.

Only the econo-car engine ran flawlessly but that car lost its gearbox oil and all cogs except first, crawling home 2nd in class to a 1.6 Carrera although it still travelled further than the 2 liter class winner. No better than 11-12 overall put a final end to any dreams of stealing that 3 liter makers' crown with little over half the displacement.

As consolation they had an outstanding hillclimb season as two semi-private Spyders held off the Maserati might for a crown again open to 2 liter cars. Taking this title seriously Porsche juggled 1.6 and 1.7 liter works engines between Swiss Heini Walter and Sepp Greger of Germany to be sure that these two would finish 1-2 in the title chase.

All those private Spyders scattered around the world continued to win as well of course: anything from a 1-2-3 in the Canadian Carling 300 to 2nd behind a D-Jag in Angola.

This RS60 was not only quick and even durable if run in its proper class, but it was also ubiquitous. There was little reason to make any but detail changes for 1961 although they did give the model a new name. Once again: the ever-willing Spyder was taking second place to their formula forays.

RS 61

It wouldn't be entirely fair to dismiss the RS61 as merely more of the same yet we must admit that there was less reason than usual to give it a new title—assuming we mean only the 1961 customer cars. The factory had several changes in mind including a longer chassis and full two-liter engines but these inevitably applied to fewer cars.

As for catalog cars; chassis numbers indicate that a further 14 of this last production Spyder were built, cars differing from the RS60 only in internal subtleties and a very slightly stubbier profile. The Spyder was already wide for its wheelbase and thus blessed with a roomy cockpit, lightweight seats which could be adjusted, three-spoke plastic wheel, the most basic dials and push-pull switches. There was even a flimsy but workable fly-off emergency brake.

These came as close to true production racers as the era would see yet proved capable of beating machines twice their capacity on the right day and track. In America this line which culminated in the RS61 literally launched a generation of stars like Penske, Holbert the elder, Sesslar, Donner and more.

To the end such cars were built by a small, elite cadre of master mechanics; one a specialist in wiring, another in engines or transmissions. Cars built over the winter of 1960/61 were parcelled out to an equally elite list of drivers who could hardly wait. Spyders were becoming so scarce many privateers turned to hot Carreras as Porsche wound down the Spyder era. There simply wasn't that much room left for pure private race cars.

Sebring opened their season again but wasn't a vintage Porsche race—a private Spyder proved their best weapon, once scoring was sorted out, and that only came fifth, little ahead of an Abarth Carrera and poor best among seven RS60/61 machines in three classes. Yet their best RS had covered 199 laps, 3 more than the winning Porsche a year earlier. Racing was simply getting too specialized for the venerable Spyders.

Porsche reacted in various ways.

One was a pair of tunnel-tops first seen during spring pre-practice for Le Mans. These had RS frames and one-piece lift-up tails but proper lids with recessed rear windows and no rear quarter panes. Sporting disc brakes and 1.7 liter engines they were called 1962 GT prototypes although their chassis numbers fell into a limbo back between RSK and

In 1961, the FIA required a "suitcase space" and Porsche provided it at the rear of the RS61. Engine access looks somewhat decent even without lift-up tail. Gurney/Bonnier (No. 134) RS61 at the 1961 Targa Florio. They finished 2nd in this 1987cc version. G. Hill/Moss Camoradi RS61 led the 1961 Targa until seven kilometers from the finish when the ring and pinion stripped. Sloniger and Coltrin photos.

RS60 so we must believe they were trial balloons as much as serious GT projections.

There was little real fuss when the French refused to consider them as future GT cars. Porsche was swamped anyway in 1961, the first year of a 1.5 liter F1 formula, and looking for sports car aid.

The US Camoradi team announced its deal to run a semi-works RS61 in all championship events except Le Mans with Stirling Moss and Graham Hill driving. If Camoradi won it was a Porsche, if not a totally private effort. This same dodge appeared at the Targa where two new "works" cars were entered by Scuderia Venezia although mechanics and team manager were pure Porsche.

These were also the first full two liter fours from Porsche, mounted in special RS61 chassis stretched to a 92-inch wheelbase which had probably been designed to take their flat eight since that's how they ended life.

The two liter, four-cam engine paralleled the Carrera 2 road cars. As a powerplant it went back to 1956 with works number 587. Among all the bore changes at Porsche this was the only four-cam engine with any but a 66mm stroke. It had 74mm and plain bearings. On paper 165 hp was hardly more than a well-developed 1700 gave but torque was improved considerably.

Bodywork looked very like those Le Mans practice prototypes with-

Moss/Hill Camoradi car (No. 20), with old style headrest (in effect, a customer RS61 despite the obvious differences), was at least as quick as factory cars and they could decide engine options without consulting factory engineers. Gurney at the Nurburg Ring (21). Despite body roll, the outside rear wheel is nearly perpendicular to ground. Sloniger photos.

out tops, but the cars weren't fully sorted. Moss/Hill even practiced with a 1.7 liter engine then started the two liter and led the race until barely two miles from the checkered flag when their new-found torque proved a little too much for the differential. Stretched bolts let the oil out when a third straight Targa victory was literally within sight.

The remaining two liter Spyder eventually eased home 2nd despite brake fade, barely ahead of its 1.7 stablemate.

Knowing they would never beat Ferrari's twelves at Le Mans Porsche yearned to win their own German 1000 km event. They left all GT worries to a fleet of private Carreras and bet all on the new 2 liter offshoot although drivers were less convinced. In the end Moss exercised his Camoradi prerogatives and swapped back to a 1.7 engine for the race.

He even led in the car when weather turned foul but retired with engine trouble and the race ended with another Italian win—Porsche had tried too hard again at home. The two liter GT class even fell to an Elva.

By Le Mans the marque title was Ferrari's so Porsche raced for class honors although they did push a two liter into 5th overall after starting four different engine sizes. Lidded Spyders came in 1.6 and 1.7 capacities, the open cars had 2 liters and there was even a 1500. Lap times indicate they had fitted very tame engines indeed, driving to arrive after losing ten cars out of ten the two previous years. Withall, Americans Gregory/Holbert provided Porsche's best finish in a semi-private car.

Adding insult to a less-than-superb season, Ferrari got its 2 liter hillclimb car working and took that title away from Porsche's supported drivers too.

By 1962, the works cars had their rear quarter air scoops reversed. This one also had an 8-cylinder engine. Sloniger photo.

Coupe version of RS61 was tried at Le Mans in 1961, as well as open cars (opposite page). Porsche garage at Le Mans where all cars were works entries. Car on hoist has latest rear suspension. Sloniger and Goddard photos.

In fact, it was one of the worst competition seasons Stuttgart could recall and they found themselves explaining that race boss von Hanstein had been "mis-interpreted" by an interviewer who quoted him as saying Porche would quit the race tracks after 1961.

Actually he said they wouldn't build any more private racers—these were too costly and had no further chance. The customer future at Porsche belonged to Carreras, at least in four-cylinder terms. When works Porsche two seaters appeared again they would be powered by eights of 2 liters and up.

Rival race firms had finally caught up with Porsche's four-cam 547 in its Spyder trim. If we record a 4th at Clermont-Ferrand it is only to indicate Spyder status at a secondary track after 1961.

Still, Porsche had enjoyed a far better than good run for its money by any measure from that first 550 at the end of 1953 to a car of the same basic design which could still win major internationals as late as 1960 and lead races or take honorable front places even if it didn't win outright in 1961.

In uncompromising prototype terms the 547 engine had run its course—but eight years out front is a solid monument for any race engine.

Heini Walter in a semi-works RS61 with various "loaner" engines cleaned up on the hills, and ran special track events for Swiss only, such as this one at Hockenheim. Sloniger photo.

F 2/1

Porsche quite literally eased into monoposto racing through the side door...of a two-seat Spyder with steering wheel on the left and most of its endurance race equipment in place.

Formula racing was probably the one use that neither Fuhrmann nor Ferry Porsche forsaw for their 547 engine but the German press was naturally eager to have a new silver arrow and Porsche engineers couldn't resist. At first it was merely a Spyder on Nurburg Ring where an F2 entry for the German GP was easy to arrange.

This second formula called for 1.5 liter engines and who did those better than Porsche. If that first one-race effort hadn't gone so well the whole scheme might have sunk from sight again. Instead it grew to a center-seat Spyder and then open wheel cars.

It is easy enough to suggest now that they might have been wiser in 1959 to concentrate on either formula cars or Spyders rather than both but once Porsche tried F2 it would alter their racing, perhaps for all time, and escalate it beyond late-Fifties dreams.

It began in July of 1957 when Porsche removed lights, covered the passenger seat and fitted tail fins to an RSK as a one-time effort for their home fans. Barth, newly arrived from East Germany, lapped the Ring at 10:02 and won the F2 cup of the German GP.

By 1958, the RSK had arrived with center point steering which made it easy to put the driver in the middle. Since a sports car would be heavier than a pure F2 they wisely chose Reims with only two hairpins and miles of straight for the debut. Behra in a tail-fin car with half spats on the wheels and headlights removed went forth against 14 Coopers, 3 each from Lotus and Osca and Collins' Ferrari.

Only the Italian car and Moss' Cooper could challenge the Porsche. The Cooper surged ahead out of hairpins only to watch Behra sail by on every straight—he led almost every lap past the scoring stand—and once Moss had to pit Porsche gained an easy 1 second-lap over the Ferrari to win going away.

Their 1500cc RSK had lapped virtually as fast as Fangio in the steamlined 2.5 Mercedes GP car on its debut there only four years earlier.

Barth drove the car later that year to turn 9:42 laps of the Ring (9:55 for a 1954 GP car had been wondrous) but a new boy named McLaren was doing 9:37 and won the F2 portion of the 1958 German GP, 5th

Barth, in one of the first Formula 2 Porsches, at the Nurburg Ring in 1957. Molter photo.

The first "single seater" from Porsche was an RSK, with fins and most other road equipment, for Behra to use at Reims in 1958, where he won. Goddard photo.

overall among 2.5 liter racers with the Porsche 6th. Finally, Masten Gregory got the center seater at Avus but ran behind two standard Spyders.

A proper open-wheel F2 could not be far behind. The first was built in a few weeks during the spring of 1959 using the 155 hp RSK engine in a car weighing some 1100 lbs. dry, distributed 47.6/52.4 with the driver aboard.

They spirited it to Nurburg Ring for a private effort in the early A.M. of May 4—with the proviso from Ferry that he might approve a trip to the imminent Monaco GP if they could return laps faster than 9:30. Quite naturally it rained that day, being the Ring.

May 5 was dry though and the car with new six-speed gearbox did 9:47 first lap with von Trips driving, then 9:39 and finally 9:29.8 in the three laps available. Trips was sure he could reach 9:25 with minimal chassis tuning and added modestly, "Moss would be 10 seconds faster still."

Both chassis weight and frontal area were reduced of course and the twin front trailing arms were bent back at a more acute angle. Ribbed brake drums and carriers were cast electron while the alloy body weighed only 65-odd lbs. With no wind tunnel time Hild attached wool tufts and watched from a following 356. About the only change was smaller air intakes behind the cockpit.

At this point the choice was open between battery ignition and Scintilla mags—with a mechanical fuel pump.

The rear suspension used the coils and wide-based bottom wishbone tried on a Spyder for Sebring. Drivers found the car quieter in back while front wheels had a touch of positive camber. Mirrors were originally faired into front shock mounts and fuel carried left and right of

Barth used the Behra car without fins to place 2nd in the Formula 2 section of the German GP at Nurburg Ring in 1958. Coltrin photo.

A center-seat RSK/F2 variant in 1959. Weitmann photo.

the driver....who was on his way to the Monte Carlo Grand Prix in 1959.

Ferry had told a journalist asking about his GP plans in 1953: "Porsche's name has been connected with GP cars for years and if means allow we would be interested again." When Trips lapped the Ring under 9:30 the means were found. Porsche was fully aware almost everybody saw this as a prototype for the 1500 F1 class due in 1961.

Only 16 cars would be allowed to qualify for the tight Monaco track out of 20 proper F1 machines and 8 F2 hopefuls (including a Colotti-built, Behra-backed, Porsche-powered F2 special which Maria de Filippis failed to qualify).

Three of those F2 cars elbowed their way onto the Monte Carlo starting grid with Trips the best in 12th among 16 starters. German silver contested a Grand Prix once again—for all of 1¼ laps. Trips slid into the wall outside St. Devote at the start of lap two, taking the other F2 cars with him.

Porsche went into PR shock with their formula car a writeoff; "and no time in the middle of a new race season to do another one," as their initial announcement went. One May start would be their entire '59 season.

Then came Reims on July 5—the best F2 race of the year with 23 starters and a battle-royal among the top three cars, two of which turned out to have Porsche power. Stuttgart had rushed through a new open-wheeler for Bonnier while Trips drove the center seat RSK which had gone so well there a year earlier.

(Today the official wording goes that Ferry was so impressed by those first Ring trials, he immediately authorized three cars for the season.)

The first Porsche open-wheeler, with mirrors on front suspension upper mounts. This car was fastest of three F2s to make the grid at the Monaco GP in 1959 and von Trips led the F2 contingent for 1¼ laps until he lost it at St. Devote, taking the other F2s (Lotus and Ferrari) with him. Photos—Sloniger 2, Molter 1.

After years of trailing arms and torsion bars in front (two right photos), Porsche came to Monaco in 1961 with A-arms and coils in front, but still with drum brakes. Coltrin and Molter photos.

Despite threats to quit F2 after the Monaco debacle, Porsche had a new Monoposto for Bonnier at Reims where he finished 3rd. Sloniger photo.

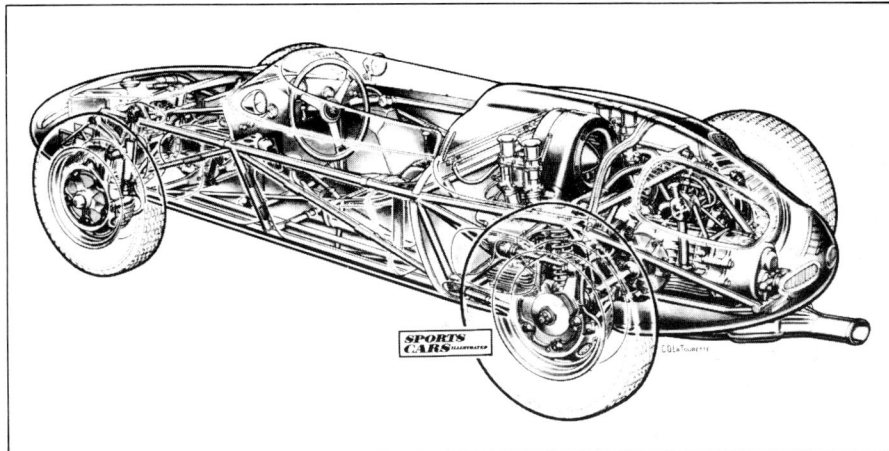

As a Formula 2 car, for 1960, the Porsche was somewhat sleeker. LaTourette drawing courtesy SCI/Car and Driver.

By the 1960 race at Solitude, Porsche was experimenting with new, cleaner, body shapes. Sloniger photo.

Still with trailing arm front suspension. The car in the background is the first of these proper F2 cars in Rob Walker colors for Stirling Moss. Werkfoto.

Porsche engines gave 155 hp then with little power below 5000. This produced a calculated 160 mph top—almost enough. The Moss Cooper had a 165 hp Borgward engine which sufficed for the win with Bonnier's works Porsche 3rd, Trips 5th. Sandwiched between Moss and Bonnier, however, was Herrmann driving the Behra Porsche which was just that bit thinner in front elevation than the factory car.

They were naturally pleased to have three engines in the top five but doubtless prompted to re-examine works streamlining. There was little further F2 racing for the works that year although private drivers ran their center-seat RSKs. Harry Blanchard started 30 seconds off the pole time in the US GP at Sebring, ahead of a TecMec, Maserati 250F and the Kurtis. He finished 7th in this 2.5 liter field, four laps off the pace.

Meanwhile the factory was testing at Hockenheim, even letting a few friends drive. One rated the car "much cleaned up since Monaco," and quite roomy although its cockpit was a snug fit compared to F1s of the time. This plump F2 was a "tiny machine" for its day.

The engine only spluttered below 5000 rpm and they still claimed no more than 160 hp with a 7800 rev peak yet the reporter insisted that six speeds were "too many gears from the human standpoint." Their pattern put 1-3-5 forward, 2-4-6 to the rear.

With rear wishbones and tighter front trailing arms the "walking" effect of swing axles was all but eliminated and the car proved very steady in Hockenheim's bumpy bends. The only body change was moving the mirrors back to the cockpit.

Brakes were particularly praised and Moss quoted as preferring their ATE drums to English discs for progressive feel and true stops. This carried weight since he would shortly receive the first of the new 1960 F2 Porsches whose enormous drums filled the 15" wheels. It was the only formula car without discs.

Barth, on the grid of the 1960 German GP, and later shown chasing the ex-Walker car (page 59), now run as a quasi-works, half-Camoradi entry. Sloniger photo.

Below. Ferrari didn't contest the German GP in 1960, run on the short, south course at Nurburg Ring to suit Porsche's F2 cars, so Trips was free to run the lower, sleeker car in miserable fog and rain....while Bonnier won in what amounted to Porsche's original machine. Sloniger photo.

Moss had tried a '59 car at Goodwood and declared, "the best way to take care of this astonishing Porsche is to drive one yourself." The ecstatic factory promptly fitted 48mm Webers to the Rob Walker car for Moss although 40mm was considered the maximum for 1500s then. They did retreat a size to 46mm Webers after one race at Syracuse.

A change in tail shape had relieved another problem—high pressure air between the carburetor ram tubes caused the air correction jet to go rich. Fitting a smaller jet before they spotted the real problem had only reduced acceleration out of slow bends.

Moss was one who disliked six speeds and the Walker car soon sported a gate to keep him away from I-II but Porsche needed them to stay within the narrow torque band. They squeezed six with synchro into a 718 box by leaving out reverse and overhanging I-II in an extension.

The ZF limited slip differential was common but here plungers slid

Herrmann in the latest body style, as raced at Monaco in 1961, with separate carburetor covers (car 6). The shape appeared again at Zandvoort (No. 9, next page), on three cars, with wishbone/coil suspension, but failed to impress. Sloniger photo.

Bare chassis of the 1961 Formula 1 version shows spaceframe, coil springs all around, and front oil radiator. Shift lever is to driver's right. Werkfoto.

sideways rather than in/out radially to make it more compact. Two ring/pinion sets were offered and two to five gear sets for each speed while the usual tires were 5.50/6.00 Continentals on 15" wheels. Wheelbase was the same as an RS60 with track reduced.

Front suspension was made as rugged as possible with sturdy threaded trunions top and bottom since the design fed in massive steering loads. The rear hub carrier was fabricated of sheet steel with the coil springs angled sharply. Engineers liked to call this car their first genuine F2, the Monaco car being its prototype, but Moss soon decided a Climax gave more power.

He took the Walker Porsche to Syracuse on March 19 and led by 29 seconds half-way through the race, setting a new F2 lap record before his engine dropped a valve. Porsche promptly blamed this on ATE who had mistakenly provided a batch to production, not race, standards.

At Zandvoort, Holland in 1961, Porsche bodywork was showing an improvement in shape, but in spite of everything the three cars entered were not impressive. Sloniger photo.

Team drivers Bonnier, and Gurney (No. 12) were back in the older cars for Reims in 1961. Sloniger photo.

Engine compartment of 1961 car, as seen at Zandvoort (as on car No. 9, opposite page). Sloniger photo.

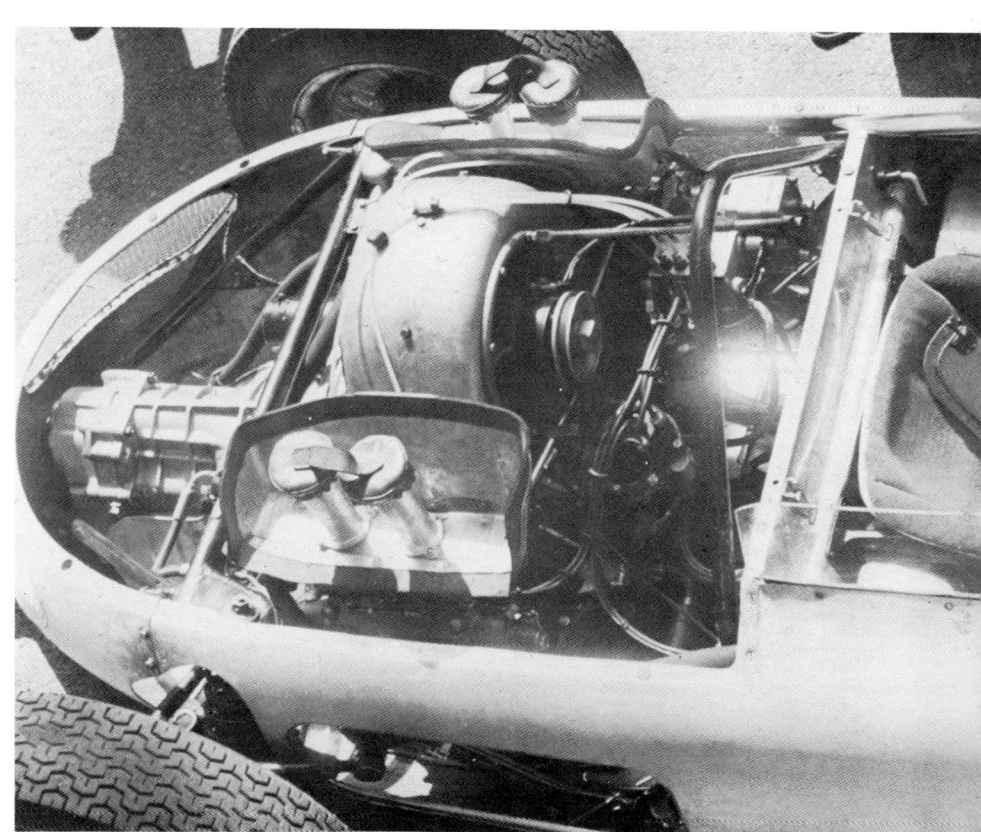

The dark blue car—actually on loan to Walker, not sold—and Bonnier's works F2 started next in Brussels where Moss took 1-3 in the heats, FTD and 2nd overall when his car jumped out of gear and spun to give Brabham's Cooper victory. Gendebien got a works car to 3rd at Pau after practicing 3rd but blamed too many shifts on a winding course for not winning.

Victory looked no closer at Aintree where a pair of Coopers led to half distance. But both retired leaving Moss ahead of Bonnier and Graham Hill in a 1-2-3 Porsche parade.

They were particularly anxious to do well on Solitude but first a Lotus led then Trips in the new rear-engined Ferrari—a bit of flattery Porsche could have done without—passed Herrmann's Porsche on sheer power. Stuttgart drums let their drivers go as deep as Ferrari discs and Gurney even had a slipperier body done by Porsche grandson Ferdinand but gear selection was dodgy and Herrmann couldn't stay with the Ferrari even when he revved the Porsche to 8700.

In the end their four cars (Surtees drove Walker's) had to settle for 2-3-4-5 which was no better than a 3-4 at the British GP.

On Nurburg Ring they ran the short, bumpy South Course for a German GP devoted to 1.5 liter F2 cars in Porsche's honor and the weather proved horrible even by Eifel mountain standards. It was a dull race in thick fog once Ferrari abstained and Porsche captured five of the first six grid positions including Trips as guest in their new shape.

At best the FIA manufacturer's cup was at stake. Cooper led on points but had to do better than third if Porsche won—which Bonnier did with the rest 2-4-5-6 and a Cooper third. Porsche claimed the F2 Constructor's Cup although that too was debated.

It was a busy if not spectacular season. Porsche took 3-4 in the Kentish 100 and Herrmann was second F2 to a Ferrari which benefited

Graf von Trips at the Nurburg Ring, shows what the combination of cornering forces and rough track surface does to wheel angles. Goddard photo.

At a non-championship Solitude F1 race, Barth drove a new car with wishbone/coil suspension, flat cooling fan. Sloniger photos.

from tows by its 2.5 teammates in the Monza GP. Without the tow Herrmann's claimed 160 hp proved equal to the Ferrari which boasted of 180.

Moss was only 4th in Denmark but headed a 1-2-3 sweep of the Austrian airport GP at Zeltweg. An Innsbruck win against second teams built confidence then Trips' Ferrari led 91 of 100 laps at Modena before failing discs let Bonnier out of his slipstream for victory. In a new year-end South African series Moss in the Scots-blue Porsche won both Cape and SA GPs with Bonnier second in each.

As 1960 wound down Porsche reputedly debated selling their five F2 cars to anybody willing to carry the banner into that new 1.5 liter GP class for 1961. Equally strong rumors said that rather than retire Porsche was building 20 F1 cars with water-cooled flat eights.

In fact it was hardly conceivable that German opinion would let them quit on the F1 threshold and the truth lay between those two tales.

Porsche and Lotus were the only F2 cars under a F1 990 lbs. weight limit and Ferry agreed to carry on with his old 160 hp cars until a new (air-cooled) eight could be readied; by the second or third race of 1961 all expected. Their chassis was considered strong enough for 250 hp if they could only find them but seven development years had pushed the four to only 160—with perhaps 10 more if jet cooling worked.

Porsche claimed there was no more power to be found below 9000 rpm and the block wouldn't stand more. Actually they would extract 190/8000 from engine 547/6 and keep the same torque, although it peaked a bit higher, before the season ended. They were spread thin in the face of 1961, already using the money other builders devoted to advertising to support their 21 man race department.

Two critical problems interlocked for 1961. Top drivers, signed for a whole season, were vital in F1 and they were hard to find when the new

Monaco, 1961. Bonnier's car (No. 2) had six cm longer wheelbase to later accept the eight cylinder engine. Kugelfischer fuel injection was also tried. Gurney (No. 4) going into and out of Station Hairpin. Molter, Goddard and Coltrin photos.

engine couldn't be relied upon and the four was obviously not as powerful as a Ferrari.

Most big teams skipped Pau to open 1961 but Bonnier and Gurney, Porsche's new warriors, trained 1-3 despite Dan's shift problems. Bonnier won heat one easily and led the second until pushed off the track. When they did get in front the fates took umbrage. At Syracuse the pair, using carburetor engines, finished 2-3 to unknown Baghetti in the new Ferrari despite wider rims for fatter Dunlops and Gurney starting from the pole. At least they had beaten the Brabham Cooper and Clark Lotus.

Porsche switched to Kugelfischer manifold injection for Monaco. It didn't boost power much but gave better torque spread. They also tried new suspension while Bonnier and Herrmann had new bodies with redesigned fuel tanks which produced air bubbles under heavy braking. Gurney's older car was best Porsche in 5th, 2 laps out, although Bonnier had been in 2nd place briefly (Porsche called this car the 787).

For Zandvoort, Gurney and Bonnier had longer 90.6" wheelbases with front wishbones and injection, Herrmann had a short wheelbase/injection and de Beaufort an injected ex-F2. Ferraris dominated that famous race where all 15 starters finished without a pit stop and Gurney was best Porsche in 10th, a lap behind. Hard to hold in fast bends the new cars with L-shaped fuel tanks in the nose at least avoided air bubbles.

They were close to pulling out of GP racing but rescued a pair of 1960 F2 machines just before they were sold and ran them at Reims which proved the thriller of the season. Ferrari's steamroller came apart with all three top cars in the pits while the Porsche pair battled Baghetti in the older Ferrari.

A race-long duel between Gurney's Porsche (12), Baghetti's Ferrari (50) and Bonnier's Porsche, ended with this cliffhanger finish at the French Grand Prix at Reims in 1961. Ex-F2 cars, like this one at right, driven by Schiller, ended their competition days in events like the Swiss Hillclimb Championship. Molter and Sloniger photos.

This trio turned its laps under one pocket handkerchief until Bonnier retired with only two laps left and Baghetti ducked out of Gurney's slipstream on the final straight to beat the Porsche by less than half a car length. Bonnier put his old car on the front row at Aintree but had locking brakes in the wet and shared mid-field with Gurney until it dried and the two could finish 5-7.

At Solitude where it mattered there were no Ferraris but Ireland's Lotus managed to edge the four works Porsches plus de Beaufort, now in his own F2 car. Porsches were faster down the straight and up the hill—Gurney got the lap record—but the Lotus put it all together better.

Stuttgart's car breakdown was interesting.

An engine with flat fan driven by vee-belt saved 6 hp and allowed a lower tail. This appeared in a long wheelbase car with the first annular disc brakes on a formula Porsche. These weighed 0.6 lb. more per wheel than their own drums but 4.4 lbs. less than English discs. Twin-piston calipers were Porsche-made to Dunlop license. Barth drove this car, obviously their future hope.

At the German GP which mattered even more Porsche still ran fours and Bonnier got onto the four-marque front grid row only to retire early with piston failure. Herrmann went from mid-grid to third but lost his clutch and Gurney in a disc-braked car was best Porsche again, 7th after a mild shunt bent his rear suspension.

All cars were back to front trailing arms and Bonnier—who had lapped within a hair of the 9:00 mark in private training—practiced with disc brakes but chose to race with drums. The flat fan wasn't seen.

Monza, 1961. Gurney (46) outlasted most of the opposition to finish 2nd overall—their best title placing—and Barth had the experimental flat-fan car again. Sloniger and Coltrin photos.

At Modena Gurney led the early laps from English fours/no Ferraris while Bonnier finally took 2nd to Moss. Porsche and de Beaufort ran old cars at Monza where the works machines and Moss' four made up a second pack, trading tows until Gurney inherited 2nd after several retirements. Stuttgart withdrew from the Oulton Gold Cup at the last minute, then Bonnier got a 3rd (1 lap behind) and de Beaufort 6th at Zeltweg (no Ferraris).

The official season ended in the US at Watkins Glen (no Ferraris) where Gurney followed a Lotus home by 5 seconds for the driver points which lifted him to a third place tie with Moss for the year while Porsche was third in Manufacturer's points to Ferrari and Lotus. The eight had yet to race.

Bonnier and Barth went to South Africa to take four 3rds, two 4ths, a 5th and 6th while machines like a local LDS Porsche came in 9th at Cape Town. This Serrurier design used an RSK engine—about what the works still used. Promising its eight for 1962 Porsche simultaneously worked on direct injection for the four which was now giving as much as 190 hp, ten more than the eight.

Dissension surfaced on the engineering staff where Tomala backed the eight while Mike May (the same Swiss who had out-winged works Spyders) put his special injection on a four and found 20 new hp. May was a GP-class driver so Porsche entered him for the Pau GP in early 1962, using his own engine. When May got to the race and discovered Tomala had cancelled his entry he left to work for Ferrari.

Porsche's annular ring discs as seen on Gurney's car at the GP di Modena in 1961. Coltrin photo. Gurney shown again at Solitude. Molter photo.

Monte Carlo, 1961, new bodywork and wheels all askew as Herrmann goes through Station Hairpin. Sloniger photo.

For the remainder of 1962 the works soldiered on with the fours, then struggled with the eight. Bonnier's ex-F2 car for Pau, Brussels and Snetterton was entered by Scuderia Venezia but serviced by works mechanics. He took a 2nd and a 3rd, more than Seidel in a weary ex-de Beaufort car or Schiller in another ex-works Porsche could manage.

Porsche only claimed 183 hp now in a new, somewhat narrower chassis, using direct injection, plus new combustion chambers and ports but the car didn't handle very well and multiple swaps between front trailing arms and wishbones didn't help.

When an eight finally appeared for the Dutch GP they loaned local boy Pon a flat-fan, coil-sprung, injected four which he spun on lap 4 while de Beaufort in his own car took a title point by finishing sixth, best Porsche and a lap ahead of the eight.

That same day second teams contested the Naples GP where Abate in the Venezia car was only 3rd to an old Ferrari and a Gilby. In the Enna GP Abate would be 3rd again but ahead of several English cars powered by the V-8 Climax.

Porsche was going to skip Monaco but finally sent an eight for Gurney who retired while Bonnier was 5th in the Venezia four, seven laps behind.

Meanwhile, de Beaufort, the epitome of a gentlemen driver, raced wherever he could get an entry and usually reached the top ten, thanks to reliability. He was 6th at Rouen for another title point—overshadowed by Gurney's win in the Porsche eight—and first four cylinder car to finish the German GP where he reputedly had a May engine. At high speed Monza de Beaufort was one of only two fours to even qualify, taking 10th in an elite field.

By the US event—where de Beaufort had his only grass-cutter—it was common talk that Porsche would leave F1 racing after 1962, "to defend its GT title." Their indefatigable Dutch count continued through the Mexican and South African events. In all he drove a private 1.5 liter Porsche in 23 F1 events during 1962, finished 22 of those and captured two title points.

The only race Porsche's eight cylinder Formula 1 car ever won (Gurney, at Rouen in 1962) overshadowed Beaufort's 6th place in the four cylinder car (38) and his second title point of the season. Sloniger photo.

Hockenheim Ring tests and Bonnier is reporting his findings to Ferry Porsche and von Hanstein. The car had the new chassis, flat fan, disc brakes and fuel injection. Kurt Worner/Road & Track photo.

The last try at extracting more horsepower from the four cylinder Formula 1 engine before the eight appeared. Worner photos.

Carel de Beaufort richly deserved the first von Trips award as best privateer in F1 racing. It was no contest. He would have won as handily in 1963 but this memorial trophy couldn't go to the same man twice so a Climax V-8 driver got it, often running behind the Dutch fours which raced much of that year in Ecurie Pan American colors.

The 1963 season opened with no official word on Porsche F1 plans but releasing Bonnier and Gurney left no doubt they were out of top rank racing.

In non-title events de Beaufort took 2nds at Syracuse and Rome, a 3rd in Austria, 4th at Pau (to Schiller's Porsche 4 in 3rd) and 6th at Enna. More to the point, the Dutchman captured two 6th places in the Belgian and US GPs, the two events where he had to replace blown engines after practice.

Those 6ths gave him two title points while his occasional backup driver, German newcomer Mitter, had three points after a fine 4th in the German GP where de Beaufort had run right behind him in 5th until a wheel rim pulled off its center. Reliability was receding as the cars got older and had to try harder.

This became all too clear in 1964. De Beaufort could only qualify 17th for his own Dutch GP and wasn't among 13 finishers. He had the only "four" at Solitude, finishing 8th in a strong field. The four cylinder GP era then came to its sad end on Nurburg Ring where Carel Godin de Beaufort crashed fatally during practice—striving to come within 10% of an ever-faster pole time so he could make just one more race in his aging Porsche.

This marked the effective end for the monoposto with a 547 engine. A few ran in hillclimbs but no more of the most interesting if least-successful four-cam Porsches were seen in Formula 1.

Carrera

"Carrera" has been written across the tail of so many overlapping Porsche road models it would be madness to try and explain them all.

Instead we'll concentrate on those fours which raced during the decade beginning in 1955. The heart of the matter was a race engine after all, the 547/1 roller bearing Spyder powerplant only slightly detuned to 100 true hp from its 1498cc. It first appeared officially (exception to be noted) just as the 356 became 356A. Thereafter true Carrera sports would always prefer the Speedster body style and later take a GT over a Deluxe when the two lines were split. Ultimately they lusted after an Abarth.

A grey Coupe named Ferdinand was really the first "Carrera" although built long before such power was dreamed of. As one of the first Stuttgart-built Porsches it was the Professor's last company car. When he died it gravitated to their test department and progressed from 1100 pushrod engine to the first experimental 100 hp installation.

Bott, of the development team, recalls; "somebody said, 'that engine has done pretty well in a Spyder, let's see how it goes in a 356.' Naturally our managers all wanted to try it. I still remember my first run—the enormous power spread which came on later than the pushrod 1500 but went so much higher. It took a little getting used to."

That same 547 very nearly ended Ferdinand's career. On their first test run in 1954, the carbs caught fire and Dr. Fuhrmann, the only one who had expected to see a four-cam engine in such surroundings from the first, had to douse his pet with snow.

But customers cried for more GT power so Porsche persevered, next sticking a 547 into one of their spare 1949 alloy coupes. Polensky/Linge went off on the Liege-Rome-Liege marathon as a "Porsche special" and won outright. The clamor from would-be Porsche pushers grew.

Of course it was no trick to keep a works GT running with a master mechanic aboard but a series of high-strung road cars for anybody rich enough would be something else. Not to mention the production headaches in even a small run of such engines. Ferry bravely approved a very small series—"maybe 50, at best 100."

These were designated 1500 GS and named Carrera in honor of the Mexican race where Porsche's Spyders had done so well. The launch coincided with their A model announcement in the fall of 1955. However a Porsche spec sheet from that same October lists a Carrera engine option for the earlier 356 Speedster which suggests they first meant the model for

A rare Carrera—as it wasn't supposed to exist. The 356 Speedster without even the Carrera script from 1955. Otherwise all Carreras were put into the 356A body. This privately-owned Carrera went to Montlhery, France, and was bored to 86mm to get into the two liter class then set records for 1,000 miles, 2,000 kilometers and 12 hours, all in the 115+ mph range with a best lap of 127.8 mph.

sport only. At least one was built but Porsche has largely forgotten this byway and offered the four cam in all A bodies instead.

They realized that stronger stabilizers, more vertical shock mounts, improved track rods, wider track, hydraulic steering damper and 15" wheels of an A gained more in handling for a Carrera than might be lost by the slightly greater weight.

The sales key in any case was that engine which would be built to full Spyder standards. Announced with 8.7:1 compression ratio it was lifted back to 9:1 for first deliveries with a rev maximum of 7000. That was enough for 125 mph with the promise of 20+ miles per gallon too. Such promises were backed up by running every engine on the bench for several hours at 4000 rpm plus several minutes at full throttle.

A Carrera Cabriolet, the heaviest and most expensive Porsche then, cost less than $5000, just about equal to their total capitalization when Porsche was founded as a car builder only six years before.

Carreras soon became the cars for actors and posers no matter how hard Porsche tried to keep them for racing. Of course enthusiasm waned that first winter when the heater-less Carrera earned its name as the most expensive icebox in Europe. With only 3000 cars of any kind to their credit by 1954 Porsche was taking a truly exclusive gamble. Yet their original "50" cars had already grown past 700 sales by 1960.

This only multiplied their service woes. The repairs foreman supposedly took one look at his first Carrera and slammed the lid—who could change eight sparkplugs he couldn't even find? Eventually 40 minutes would be considered good time for a trained man from the "Carrera" school to change all eight plugs. And who could hope to handle such performance?

In fact Carreràs weren't as mean as feared, once loyal Porsche drivers learned to run them at revs which had put their best S model over the red line.

A very early magazine road test noted that once-strong oversteer tendencies were "totally eliminated" in the 356A, which seems over-enthusiastic. Better stated: oversteer was milder and came much higher on the handling scale. The tail still came out if you lifted off and a canny Carrera man knew he was near the limit when his steering went light. That went with considerable elasticity for a 1500 of that era. This car could even be raced or rallied by less than pro drivers as long as the owner remembered not to drop below 2500 rpm once the car was run-in and learned the occult cold-start drill thoroughly.

Not mean, it was still a 1.5 which went from 0 to 60 in well under 12 seconds and on to 100 (bottom three gears) in less than 30 seconds, even with 125 mph gearing. Using short gears 0-60 times as impressive as 10.8 seconds were possible. Small wonder competition types loved them. The small tank and poor lights could be corrected.

By late 1956, entire starting grids might be Porsches and this one, at Nurburg Ring, has a front row of Carreras. Bauer photo. Carrera GS GT had carburetor cold air boxes in 1958. Werkfoto.

The Liege-Rome-Liege rally was not a Porsche victory in 1956, but Porsche won the event three times in the second half of the Fifties. Kurt Worner/Road & Track photo. Opposite page, Carreras won everywhere, including the banked Avus track in Berlin. Sloniger photo.

Then the original euphoria wore off a little and Carrera fans split into those wanting the name without the game, and serious racers who hated to haul all those frills around. The proper competition Carrera came in the model's second year when Porsche divided this 1500 GS line into Deluxe and GT.

The GT was now available only as stripped Coupe or Speedster with bucket seats, slide-up plastic windows, lighter bumpers and Spyder front brake drums since it had 110 hp like an RS. The point, of course, was that this stripped car was an eligible GT racer which outshone many production race cars of its era. Factory friends even got engines with further-polished parts, higher compression ratios and sharper cams. Racers already knew that one plug range suited all four cylinders but the front carburetors needed larger jets. Top pit mechanics could adjust valve play faster than they changed plugs.

The most coveted model was a Carrera Speedster GT with plexiglass side curtains, 21-gallon tank and "sports" exhaust which got the gasses out and noise be damned. With 7500 rpm now considered safe such a 1770 lbs. Porsche would lap Nurburg Ring in 10:45 and certainly display more than 120 hp. The cobby engine smoothed right out by 4200 rpm.

Clutch slip still followed repeated drag strip starts but the Speedster would now do 0-100 in 28.2 seconds. In fact, the clutch was their chief problem for racing, and glazed faces required a new clutch about every third race.

Testers found the brakes unfadable in 10 straight panic stops,

Strahle carried license number V 2 on his race/rally Carrera—an apt number if a hot Carrera ever had one.

however, while near-neutral handling (as understood then) came with a very clever throttle foot.

To go any faster took a works GT. As late as 1959, Porsche won a Targa class using a 130 hp roller-bearing Carrera long after customers got plain main bearing cranks. This was a Spyder with doors really, with extremely close-ratio gears (only 700 rpm between the bottom ones) and ultra-stiff shocks. It set the upper limits for those Carreras winning Sunday after Sunday around the world.

Following their first Liege victory by a "pre-Carrera" this trans-European gravel bash became something of a Carrera domain. Storez won again in 1957 with just 20 penalty points to 668 for the 300 SL in second place and a third victory came in 1959.

The Carrera matched reliability against sheer power. In the 1956 Lyon-Charbonniers rally only half of 118 starters saw the checkered. Three Carreras started, three finished and all were in the top five including 1st-2nd. When lumped with the 2 liter cars they won there too until class victories became so routine even the house magazine only bothered about high overall places.

Carreras swept class, team prize and 2-3 overall in the brutal Alpine rally, then grabbed a 2nd in the Tour de France against GT cars of twice their capacity while a private, class-winning Carrera in the 1957 Mille Miglia averaged 81.9 mph—faster than the 2 liter GT winner. Running in the 2 liter class at Reims a Carrera won the 12 hour with a time which would have credited a Spyder two years earlier.

At Montlhery a private Speedster was bored to a one-time 86mm (1529cc,) fitted with special rings from its owner's factory and run for 2 liter class records since Porsche held the 1.5 marks anyway. It duly captured endurance records for 1000 miles, 2000 kilometers and 12 hours, all in the 115+ mph range, with a best lap of 127.8 while making up time after a pit stop. This car weighed only 1690 lbs. and ran with baby VW brakes—needed only for routine pit stops.

The 1958 season was more of the same: 2-3 overall in the Baja-like Tour de Corse, overall victory in the Sestriere rally against 2.5 liter works Lancias and V-8 Fiats and just about every GT prize going from the 178 bends of Nurburg Ring to Lime Rock.

By this time only two final drive ratios were allowed so Porsche dropped the 4.86 option and kept their universal 4.428 giving 14.6 mph/1000 rpm plus a 5.17 hillclimb ratio for 12.6/1000. This was the brutal U.S. ratio for semi-race Speedster GT cars, good for 0-60 in 8.7 seconds. They also lengthened second gear while shortening third and fourth.

None of these improvements could still the capacity discussion

Factory shot of a Carrera 2 engine before installation in a car.

though. FIA class breakdowns now allowed 1.6 liters (suiting pushrod Porsches) yet the GS remained 1500. After all, they had now built 10,000 cars but owners from race champs to Prince Bernhard would trade pushrods for a four cam whether it had only 1.5 liters or not.

Porsche was well aware of the capacity problem and had the answer but it amounted to replacing one of the most successful engines of all time with a new idea. When the title did become 1600 GS (still in the 356A body) there was no more roller bearing crank. Engines 692/1,2,3 and 3A, just like 587 which was designed first but followed in production, had four-cam heads but plain bearings for rods and mains—and that took

Sepp Greger, the regular GT hillclimb winner in his Carrera on Schauinsland in 1960. Sloniger photo.

some adjustment for the Carrera fan.

This isn't to say that various racing Carreras didn't appear with 1.6 roller bearing engines but officially it was plain for 1.6 liters, rollers for 1.5 liters. It merely took race scrutineers several months to realize that all those roller-bearing 1600s weren't quite legal. On the other hand the plain bearing engine worked as well, moved the Carrera closer to production Porsches for some service work and required less care.

The separation of sybarite and sportsman became even clearer with the 1600 GS Deluxe with 105 hp and 2070 lbs. versus the GT weighing only 1860 lbs. but producing 115 hp. Porsche blandly continued to quote precisely the same top speed of 125 mph—which looked even better as a round 200 km/h—and would do so for all Carreras.

Klaus von Rucker, the design boss then, plainly hated roller bearing cranks and once they had race-proven plain bearing cranks he hurried them into the Carrera as a 1600, saving the available 2 liter version a while longer. The engine did require more oil cooling now, particularly for racing, so they added 33 feet of tubing to the 692-powered cars, carrying 10.5 qt. of oil to a pair of coolers under the headlights and back. A thermostat controlled all flow.

To prove that plain bearings could take it factory men often went to 8000 rpm in the gears while posting 0-60 times as low as 10.8 compared to about 12 seconds if you observed the red line. Considerable clutch slip was required for hot starts.

Further aids to the sporting set were a new gearbox (all Porsche) with shorter lever movements and more precision; also Koni shocks. The GT was marked by cold air boxes built into the engine lid to fit down over their dual carb throats.

The capacity increase came in part to make up for more restrictive mufflers under new decibel laws but their so-called Sebring option returned 10-15 lost hp to the track machines. A "simplified" GT interior amounted to no sound deadening, no window winders, alloy doors and lids plus the plexiglass windows—all to save some 145 lbs. despite the weight of a larger tank which filled the nose.

The single-leaf, transverse, camber-compensating spring which could be fitted to a Carrera reduced oversteer in fast bends so you could drift one with an educated right foot. Those new radial tires were another boon to Carrera hustlers.

Oil cooler was squeezed into left rear fender on Carreras. Poole photo.

Put together this led to 1959 highlights like their third Liege victory taken by Strahle in his famous race/rally Carrera carrying license V2, an apt number if ever a hot Carrera had one.

A works car in the Nurburg 1000 km that year was also the first Porsche trying discs in public but von Rucker was not impressed, growling that pads were gone in only 27 laps (barely 400 miles) and besides they weighed nearly 9 lbs. more per wheel than their own drums which would continue unless/until Porsche designed a disc which proved superior. One gathered he wasn't hopeful.

If it weren't for competition the Carrera tale would have to end temporarily with 1959 when the 356B was introduced without a four-cam in the catalog. Their S90 was supposed to thrill the Carrera Deluxe crowd. In fact Porsche soon built a series of some 40 lightweight B-Coupes for 1960, using 1600 GS GT engines.

These tend to be overlooked in light of a score of Abarth Carreras at the same time but at least they did keep the name alive in a B body while customers with winning records could have extra hp in their light B, even beyond the exhaust option ordered.

Expecting few to drive to the track any more the options became System I—no mufflers—with up to 128 hp/6700 from the best engines or System II (Sebring) equally unmuffled—for 135/7400. The real difference lay in whether you wanted a rally power band or pure peak track revs.

Such 1.6 Carreras could still take 1-2 in the GT class on high-speed Montlhery but they were beginning to run into lightweights like the Lotus Elite which made something more than a dieted B necessary. It was the Abarth, an Italo-Teuton Porsche people never wholly loved, even if it was quicker. At the 3-hour Goodwood TT, for instance, an Abarth beat a light B in class with the pair 4-5 overall, pretty well summing up the two variations on the B chassis.

One thing your Carrera could do that no Elite dare try was tackle tabletop English tracks one day and go for an Alpine cup the next. Still, England remained about the only land where Carrera ownership didn't guarantee GT laurels and even that small blot always rankled a little.

A few four-cams did race in England—one 1960 model remained competitive until 1963 without ever winning a big one, then passed a magazine "used car test" with nary creak nor rattle. It was loud and stiff of course and didn't have much punch below 4000 rpm but how many of

Carrera instrument panel. Turner photo. Another Greger Carrera on a hill, in 1964, when he could run without bumpers. Sloniger photo.

Proper Carrera exhaust setup. A 1600 with compensating spring.

Typical race finish for Bruce Jennings, who won more races in a Carrera than any driver in history. Right. If you don't like the way it runs, take it out and fix it (at the Eberbach hillclimb in 1962). Worner photo.

its 1960 peers were even mobile then, let alone capable of both 16.2 second quarters and 100 mph at only 5000 rpm.?

That sums up the whole Carrera mystique of road and race as we approach the final Carrera, 2000 GS which posted the odd win of its own, even though it had the least-known competition record. A Carrera 2 cost nearly as much as an Abarth and carried much more weight but it did revive the name publicly since the world had largely ignored those lightweight Bs.

This 2000 GS began as a 356B too, the fastest road 356 ever built although its quoted top speed remained 125. Very shortly after actual sales began in April 1962 (it was announced at the Frankfurt show in September 1961) it was also the car which introduced disc brakes to the general Porsche public.

These brakes were the most confusing feature of Carrera 2 history. Originally announced as a B with drums it then became a road test bed for their own annular discs but was handed around for first press tests in early 1962 with drums after all. It stopped well that way but the pedal was heavy. Eventual customer cars had Porsche-Dunlop ring discs.

It remained the only road Porsche to use them. And it appears that a certain number of later Carrera 2s, (in the 356C body) used conventional discs although some may have been conversions. Teutonic thoroughness has never extended to Porsche record-keeping.

These "race-proven" discs had the caliper inside the disc for greater leverage, reducing servo need, and smaller brake cylinders. Thus the Porsche design was lighter. Also, they weren't copying some foreign sports car.

Once they found this workable disc of their own it was fitted to all race and formula cars as well as to Carrera 2 although racers used an alloy rather than iron caliper. German testers promptly rated them the finest brakes ever.

While they did stop a car short, extreme heat flowed outward to the securing bolts around the periphery of the disc ring and this began to cause fractures around the bolt holes. Development costs soared and the design finally proved too costly to perfect.

Carreras go to Concours, too. This beautiful Carrera 2 Speedster was at the 1976 Porsche Parade. Turner photos.

From the standpoint of handling at race speeds a Carrera 2 was still short in the wheelbase, with even more weight in its tail, yet they tuned them to understeer until you neared the limit. Porsche also toyed with a limited slip differential but confined it to track cars.

The 1966cc plain-bearing four was number 587, first designed before Fuhrmann left the firm in 1956, and the first Porsche to use any but a 66mm stroke. With 74mm the Carrera 2 had a higher mean piston speed than its predecessors and that fabled 125 mph required pistons to reach 53.5 ft/s. On the other hand torque remained 94 lbs-ft or better right through their green tach zone from 2200 to 6200 rpm and such a spread (41-117 mph in top) made this a more easily mastered club racer.

The 587 was also a bit heavier and nearly an inch wider so that tales abounded of race teams hiring an asbestos octopus for plug changes. Access panels in the fender wells aided in the removal of the plugs.

Viewing that 130 hp customers and journalists alike wondered loudly why they couldn't have a 2 liter GT Carrera—alongside those rare, converted Abarths. There was a "GT" option for the 356C line generally with the usual alloy doors, plastic windows on straps, close-ratios gears and limited slip but Porsche "can't recall" that any of them had Carrera engines.

Outside evidence points to a small series with a 160 hp hop-up but the proper competition purpose of the Carrera 2 was to legitimize a two-liter Abarth offshoot for real GT efforts. They had to sell a series of the base model before converting the specials.

In its own right a Carrera 2 won the 1963 Midnight Sun Rally, that Swedish event where one speed stage consisted of climbing up a mine shaft. And it provided the chassis (in 356B form) for the wildest pair of Carreras ever.

But there was really too much plush about the Carrera 2—the ultimate 356—to turn it into a true track star. Also, racing had gotten too specialized for the dual-purpose GT ideal which had both spawned this Carrera line to begin with and was then so richly served by shoals of those same four camshaft Porsche road racers.

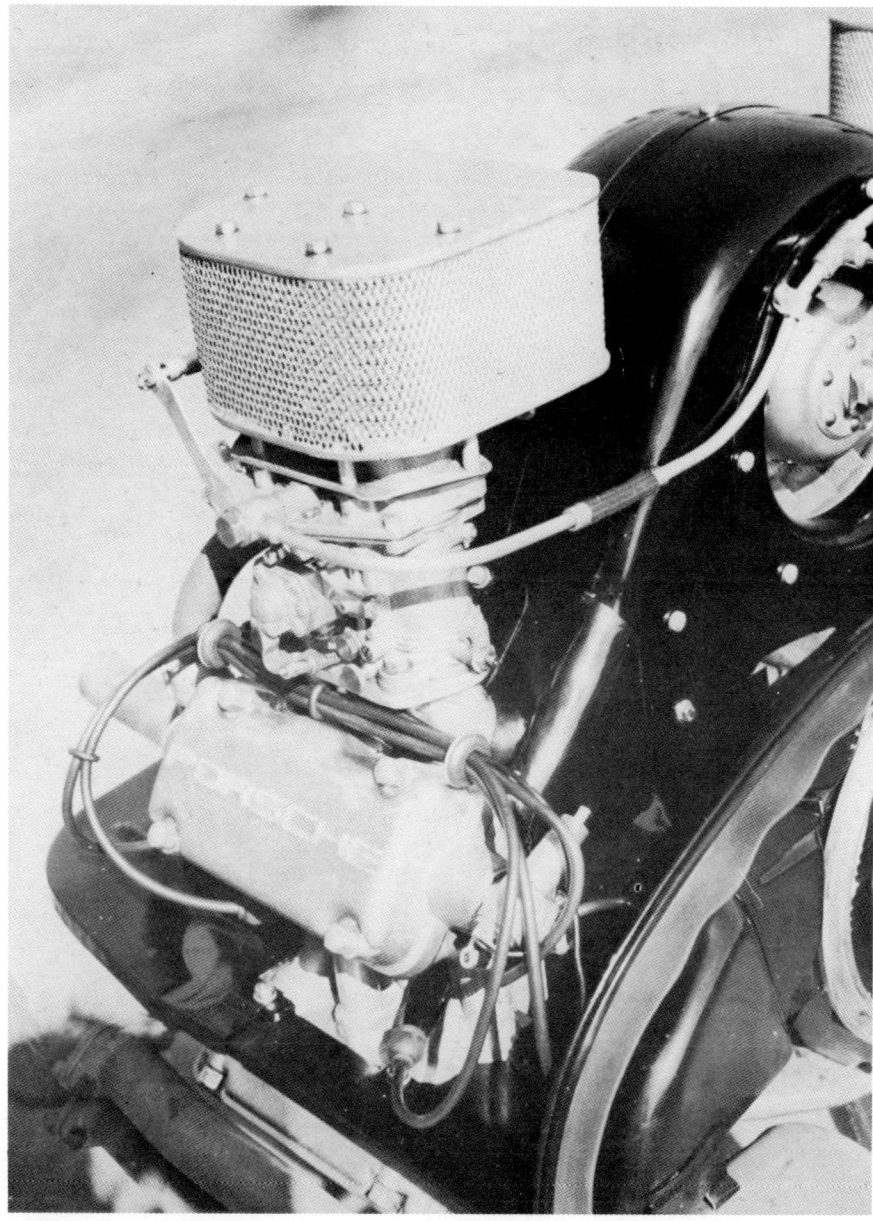

Another view, this one from the opposite side, of the Carrera 2 engine eight pages back. Werkfoto.

Abarth Carrera

There was never any doubt that the Abarth Carrera was built for competition. In appearance this GTL formed a sort of parenthetical comment between the last 1600 356Bs and the Carrera 2. Chronologically as well as mechanically the Abarth was a bridge between the 1600 and 2000 GS lines. No more than 20 or 21 were ever built (sources will probably never agree) during 1960/61 when there were no other official Carreras except those few lightweight B specials.

Many were later fitted with the two-liter four-cam engine as well. With barely a score of the GTL shells available Porsche racers had to recycle them. Repeatedly.

The car was really an anomaly in Carrera history—the one Stuttgart let outsiders tailor. In this case, Italy, where they at least had sound ties. Carlo Abarth had been involved with Porsche's Cisitalia project as an Austro-Italian who could operate in both lands. By 1960 he had become that man who fit very wind-cheating bodies onto small-engined chassis and made them go indecently fast. Porsche decided to let him try with a Carrera B.

It would seem there were 21 such chassis numbers but it is possible that as few as 18 of these were actually Abarthed. Confusion stems from the habit of converting and updating them—perhaps a new nose for long-distance lights or even a road rebuild after racing. What looks like two cars in photos may easily be one chassis.

Even the reasoning behind the model is debated. The GTL was built: either because the lightweight B wasn't quite enough to win or more probably, that B special was built because Abarth couldn't do more than 20 GTLs and those sold out so fast Porsche produced the B-GT to sooth disappointed customers. Impeccable sources support both views.

The first GTL to arrive from Italy was emphatically not up to Porsche standards, not even for track work. Nordic sizes at the Stuttgart factory couldn't even squeeze inside. They soon reworked it and eventually kept 4 or 5 on the works strength through 1961 simply because they were the fastest Carreras ever built, despite all flaws, and that was the point after all. The FIA kindly considered it already homologated as merely another GS body variant.

Their shape came closer to a scaled-up Fiat Abarth than a 356. Looking sleek it was generally called the lightest Carrera, with leather straps to hoist plastic side windows, roll bar but no bumpers, thin,

Two Abarth Carreras at Goodwood, England. Differences between No. 23 and first car (previous page) can be seen in quarter windows, taillight location, louvers, license plate lights and lower removable body panel at rear. Goddard photos. The interior is an Abarth factory photo showing deluxe interior.

recessed door handles and engine lid air scoop opened by the driver as necessary. He sat in a skeletal bucket and contemplated a 10,000 rpm, telltale Spyder tach.

After saying that—it turns out they were really only a little lighter after all (more window area). But the center of gravity was lower, the nose fell away so drivers could pick their clipping points with more confidence and they were certainly slipperier.

Even though a GTL cost over $6200 it would do an honest 138 mph through the speed trap on Mulsanne Straight and Linge, works racer/mechanic par excellence, is sure the car had another 5 mph by the end of Le Mans' long road.

Like most Porsches, initial acceleration from rest was poor. The rationale was that you only had to make one fast start per race. Pickup in the gears was exceptional for a 1740 lb. car of only 1.6 liters. There wasn't much action below 5000 rpm but you could take one to 7800 regularly—on plain bearings at that.

Incidentally, one early Abarth was a 1700, used by Walter in the 1961 2 liter hillclimb chase.

The original 1.6 Abarth came with three engine options, lumped as 692/3A. Power depended on exhaust. Decibel rules had been tightened, although lenient by modern standards, and Porsche's tamest 115 hp car could barely pass with the "stock" exhaust. Two stages of race exhaust gave the same outputs as a B-GT.

Their hot setup was dubbed a Sebring extractor to honor their prime market and in memory of a class win there although the first Abarth victory came even earlier, in the car's very first race, the 1960 Targa Florio. Porsche won its class at Le Mans that year as well, getting the new GT off to a proper start.

At the time of the announcement—they were all sold anyway—Porsche offered only the "standard" final drive "on principle" but three full gear sets for the new 741 gearbox with two more in preparation. Those same principles (translated as product expediency) dictated a single Abarth Carrera color, ex works; metallic silver with black leatherette interior.

In fact, several of these GTLs were even delivered with proper roll-up windows. Later models featured smaller rear quarter windows too, with more rounded corners to stop a fatigue crack problem. Forty-eight engine lid louvers remained the norm while the "usual" nose had three slots—a center one for the oil cooler flanked by two for brake air.

A special Abarth appeared in the Nurburg Ring 1000 km in 1960 and ran as a prototype because it carried Porsche's own annular discs. The car stopped no shorter than drum-braked Porsches and finished 7th overall.

The garage at Le Mans in 1961. A Carrera Abarth is having serious engine work done. Sloniger photo.

Recent photos, right, of Dean Watts' restored Carrera Abarth. These are not only good racing cars, but can be driven on the road or shown at a Concours d'Elegance with equal success. DB photos.

One of the better "private" Porsche pushers, Ben Pon at a GT race accompanying the Dutch GP at Zandvoort, left. Below, Abarths in private hands were still running strong in 1962 (here at Le Mans). Sloniger photos.

In 1961 another disc-braked GTL running with the prototypes took 8th—just behind a pair of proper GT Abarths.

Even customer 1600 GTLs could lap the Ring in 10:23 during 1960, 20 seconds under the class record, and an Abarth Carrera later became first GT to break the magic 10 minute barrier on this 14 mile mountain track, doing 9:58; only just over a minute slower than a good 1.5 liter GP time.

A few "road tests" appeared during 1960/61, mostly short blasts up the nearby autobahn plus a trickle through town in the 2500-4000 rpm band since a Sebring exhaust put out 103 dB at full cry. An Abarth took such low revs without a stumble but to promote GTL-worthy acceleration times like 0-60 in 8.7 seconds (ratios not specified) or 100 in just over 20 seconds you had to slip the clutch badly.

Handling was a matter of point and step down—once Porsche sorted out the original Scaglione-built body which had near-zero steering lock. That first Abarth (which they kept at the works since it was the lightest of them all) was also noted for shipping water and keeping it aboard, in the seat and footwells.

All Abarths began with drum brakes and those later receiving discs used Porsche's own. Many also received the two-liter engine, giving Porsche racers a car to remain in contention until the 904 was ready.

First entries with 1966cc—Porsche homologated them in the two liter GT class too—came at the 1963 Daytona races. On Saturday von Hanstein won the 2 liter class, 7th overall, at 122 mph in a 250 mile race using only the banked oval. In the 3-hour Sunday race over Daytona's road course, counting for the GT title, Bonnier was two liter winner from Holbert. The pair finished 6th and 7th overall, a private Abarth 8th.

This GTL rapidly became Porsche's maid of all racing in the early Sixties while engineers and designers were busy on their formula program. They even sent Barth to the German Rossfeld Hillclimb in an Abarth as a Ferrari shadow and lucked out again. GT cars raced in the dry, proper race prototypes, running later in the day, had to go up the hill in the rain. In the record books, a 2 liter Abarth Carrera won overall.

Small wonder drivers coveted them. The GTL conferred instant GT status; it was durable, versatile, and it won.

2000 GS/GT

On Porsche's family tree the rarest Carreras of them all pre-dated the 356C but paralleled the original 356B Carrera 2 which provided their foundation.

This matched set of extra-special Carreras looked far more like those tunnel-top Le Mans cars of 1961 which had not been proper GT machines at all but lidded Spyders. To identify them look for the body dividing line ahead of the rear wheel arches which indicates the flip-top tail. They had no rear quarter windows either.

To make their 2000 GS/GT Porsche kept that profile but fitted small windows either side behind the doors and put it all on regular 356B chassis carrying Carrera 2 engines of suitable output. These had normal tail hatches, not the one-piece kind. Furthermore the chassis numbers indicate they had been around for some time before the public bow.

In Porsche racing circles the model was nicknamed *"Dreikantschaber"*—which translates literally as triangular scraper and pretty well describes the form. These were blatant loophole fillers for a two liter class which only required that chassis and engine be built in a series of 100. Bodywork was free and Porsche had the shape handy.

As Bott put it, "in those days the Spyders and the road cars were not so very far apart anyway." Thus this 2000 GS/GT could be called a "road" car, look like a pure racer and boast anything form 155 to 180 hp on 1966cc, depending on race distance. They were works specials on the far fringe of Carrera history.

Their working life was almost totally competitive yet many sources had already called that original lidded Spyder "the new Carrera" and even though Porsche had no production plans when this GT version appeared many hoped it was the Abarth update. The "scraper" was drafty, unheated and leaked at every seam which makes it less charitable for Porsche to complain about Abarth's Italian body quality.

At Sebring in 1963, the GS/GT duo took 9th and 10th overall. Then they finished 3rd and 4th overall and won their class in the Targa Florio with four Porsche-Dunlop ring discs for stopping. On Nurburg Ring a GS/GT was even first Porsche to finish, 4th overall, after their two-seater eights failed and the crowd went home in disgust.

Banking on such proven sturdiness Porsche sent the twins to Le Mans to back up the eights again, running one with discs and the other with drum brakes. The score was one overrevved and the other out with a

Rarest Carrera of them all—the B Chassis 2000 GS/GT. Only two were built. Modeled after Le Mans prototypes, the aluminum specials were considered, but rejected, as customer GT specials. This one (66) is at Solitude.

A 2000 GS/GT in the paddock at the 1963 Targa Florio. Driven by Barth/Linge, it finished 3rd behind the 2nd place Ferrari of Scarfiotti/Bandini/Mairesse, and the 1st place 8-cylinder Porsche of Bonnier/Abate. DB photo.

The 2000 GS/GT was nicknamed *Dreikantschaber* which translates literally as "triangular scraper" and pretty well describes the form. No. 31 is the Walter/Pon "scraper" at the Nurburg Ring in 1963. Sloniger and Worner photos.

broken valve. Despite Mulsanne trap speeds of 143 and 149 mph to squash the myth that an Abarth was fastest Carrera after all, Porsche had missed the premier GT prize in racing with a model which did so well otherwise.

During the 1963 hillclimb season one GS/GT booked 100 points as best GT over the nine events while the runner-up make earned only 48 points.

Stuttgart started the 2000 GS/GT at Daytona in early 1964, now considering its Abarth a "production" GT. They took home GT points from Sebring with 11th and 12th overall but the best "scraper" finished two positions behind the first 904, still a prototype then, and that meant the GS/GT had run its short but rapid course.

As Bott put it dryly; they decided such an alloy version of the Carrera 2 would be too complicated and fragile for even selected customers to race with any consistency. So the two works cars remained the only GS/GTs ever built, as well as truly the last 356-based, four-cam offshoots.

Porsche turned its attention to plastic racers and thus to their own young house designer, a third generation Porsche who was ready with a brand new kind of Carrera.

904 GTS

This Porsche 904—Carrera GTS if you prefer—proved to be the ultimate four-cam car to end a line which spanned ten years.

Although called Carrera and built as a closed car its 180 hp race engine (there was milder road version but few noticed) plus stark trim set this car much closer to the original Spyder concept. It was the one Carrera with no body or chassis ties to the 356 line whatsoever—but the engine remained familiar.

The 904 appeared at the end of 1963 as a mild surprise since Porsche had just announced its 901 (later 911) with six cylinders and that seemed to set a period on the four-cylinder era. But they badly needed a new GT weapon after two years of less than lustrous racing and they needed it fast, to open the '64 season.

The answer was GTS, pet of Ferdinand "Butzi" Porsche, head stylist in his grandfather's firm. This particular Porsche was convinced that sports cars should have two seats. Period. In terms of GT success he was certainly right, providing a C_x value of only 0.34 and height of barely 42" with space for all the aggregates.

A wooden buck of his body was rushed to Heinkel who began the plastic shell in February of 1963, sending the first body to Porsche for a proper car in November. And since their six was hardly race-ready then Stuttgart worked over that faithful four to power just one more car.

Ferry Porsche had admitted in mid-1963, that a new weapon was coming and when Linge turned 9:30 laps of Nurburg Ring the 904 had arrived. They would have to build a fast 100 to qualify for GT events but orders easily exceeded that figure and far more than the eventual 106 could have been sold easily.

This 904 abandoned the Porsche mainstream in several ways. For one Heinkel, rather than the race shop, built and wired them. More basically, it was the first proper plastic Porsche although they had some experience with fiberglass 356 panels and F1 body sections.

Once out of effete formula racing Porsche went down the opposite road with a vengence. Their 904 was no fragile sprinter but a robust, reliable car built as much to outlast as outrun the opposition. They got the first 100 out by April 1964, giving customers a GT weapon again. The factory could race theirs in four, six, or eight-cylinder form to suit. Chassis 1 through 6 were works cars, identical to the next 100 for customers.

Extra lights were necessary in the Tour de France, to the detriment of aerodynamics. Weitmann photo.

Left front (from front) and left rear (from rear) suspension of the 904 shows direct result of Porsche's track experience with Formula 1 and 2.

 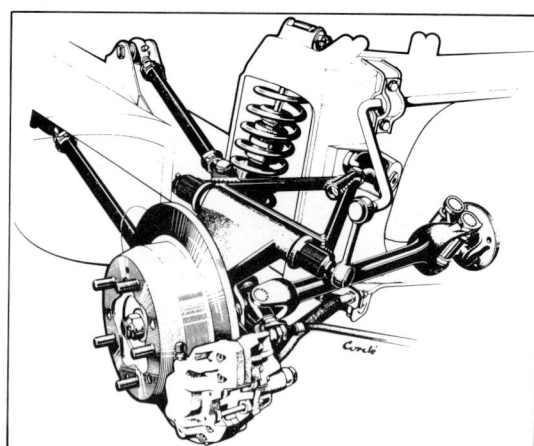

In 1963 Porsche turned to its first fiberglass GT, the 904, which was built for rough road events, but was still a pure racer, and the ultimate 4-cam car.

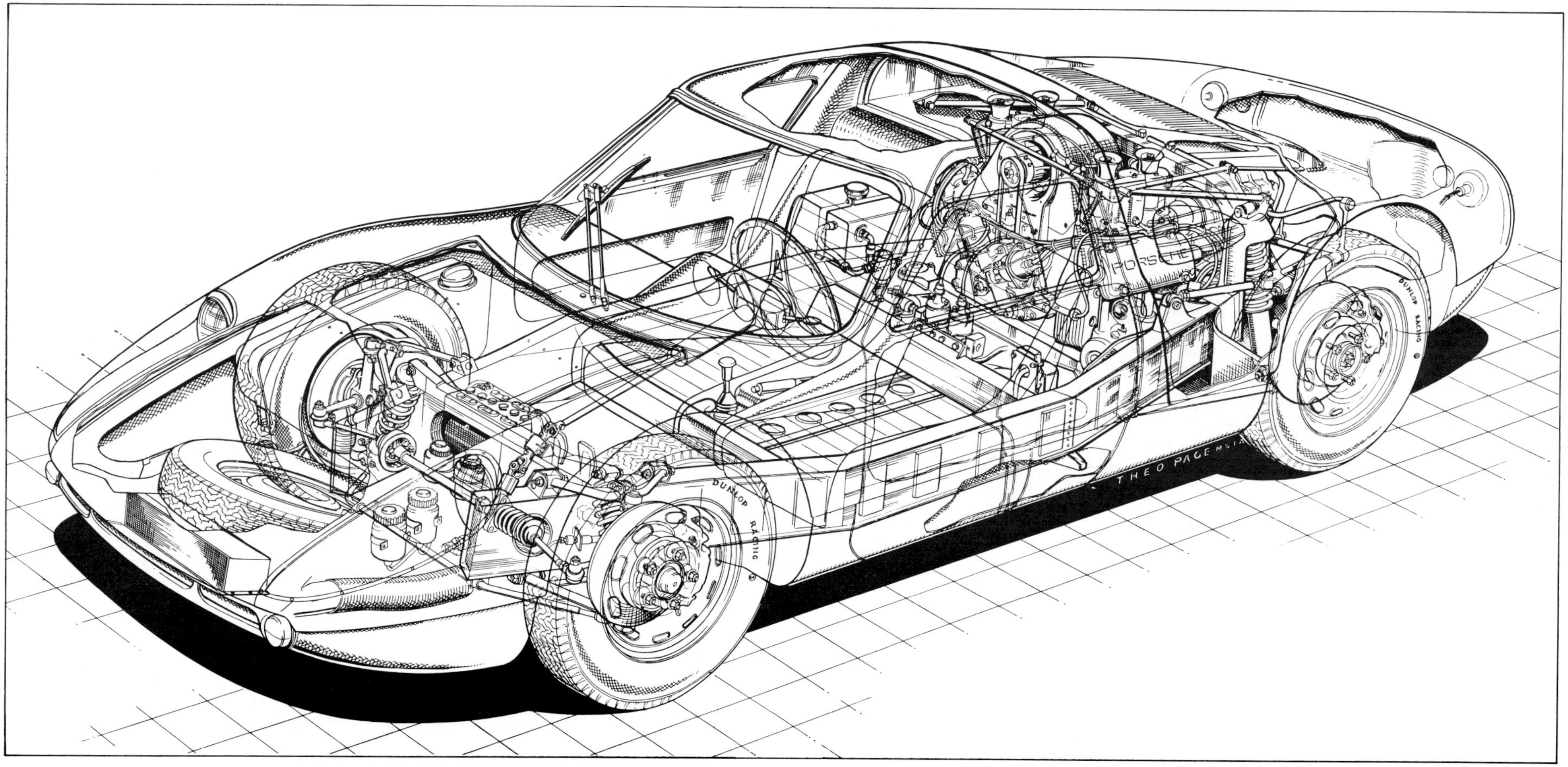

The 904 chassis was a deep-section box affair to which the body was bonded, making rust hard to get at.

A 904 at the Heinkel works, in front of a Potez-Heinkel CM 191 jet. Weitmann photos.

Plastic was chosen because it was more practical than thin alloy in lots of 4-5 a day. It was a new game then—these were the first German plastic cars in any numbers. Flex was an early problem. A test machine bent as much as 14% but they discovered that further curing of the body cut this figure in half.

Underneath the body their engineers designed in considerable stiffness using deep box-member frame rails and bonding the body directly to them. These light but beefy members bowed outwards between the wheels to embrace the passengers and engine which rode ahead of the rear axle line in proper racing manner.

A five-speed gearbox, borrowed from the new 911, hung out behind. In all the frame weighed only 100 lbs., the body another 245. A lift-up tail made this the first Carrera with proper engine access. Suspension came right out of their formula cars: coil springs with twin transverse links in front, modified wishbones in back. Brakes were four conventional discs, a first for racing Porsches. They fitted 15" wheels since smaller ones wouldn't suit tough going.

Two fixed bucket seats were set in a spartan interior. Side windows were sliding plexiglass, sound proofing near-nil. A single pantograph wiper stuck to the glass at speed but left the corners wet. On the other hand their shape flung most water up and over.

Engine work amounted to fairly extreme valve overlap and hotter cam grind, new piston profile, wider cylinder fins, twin-choke 46 Webers and generous valves. Even more impressive than that 180 hp was the flat torque curve and relatively wide torque band.

This was the first Carrera with one-digit weight/power ratio—8 lbs./hp. It was put together to prove Porsche still had the touch for a sports car surprise.

Although neither as light nor as aerodynamic as first hoped it would still turn sub-6 second 0-60 times and do 160 mph in top yet cost only $7500; cheap for a winner. Old Porsche hands found it relatively comfortable, fun and easy to drive fast.

Porsche 904s were seen at all major circuits, such as Le Mans (32), Nurburg Ring (72), Reims (45) and the Targa Florio where Davis and Pucci (86) finished 1st. Photos by Goddard and Coltrin. Interior of a Concours-prepared 904 at the Porsche Parade. Turner photo.

They were never quite sure whether to let a 904 out on test, however. There was supposed to be a road version although most eventual daily users either fitted a six or went the gritted-teeth route and kept 180 hp for its panache.

By any Porsche standard the 904 was great to drive in the wet—apart from water which still got inside since there were no door seals. If it couldn't entirely outpace rivals like the Abarth 2000 GT it handled better and lasted longer.

Entry was acrobatic and drivers over six feet tall found it tight even if the roof liner was one cloth layer thick. A large right foot was needed to brake hard but their new gearbox was good, even for a Porsche. Some disliked the pattern with low left-back and the rest in an H-pattern of course.

Purposeful and functional are the adjectives for young Porsche's design which used all available inches for machinery plus the mandatory FIA suitcase bin and large fuel tank. In road terms the low-slung car would carry a couple of kit bags—and probably drag its belly pan to do so.

The 180 hp engine was lumpy at idle (around 1500 rpm.) You could run one in traffic but it would consume respectable quantities of oil, plugs, exhaust valves and cylinder head studs that way. Grand Touring might be the name of the game in 1964 but Porsche didn't mean for anybody to take the title literally.

The first dozen customer cars were rushed to America where racing began earlier but the SCCA didn't consider it a production car right off and no 904 could hope to match the flyweights in sprints. Its field was trans-European competition where arriving could be at least as tough as winning. At that game their success story covered several seasons with the important ones 1964/65 when the GTS was fresh.

On press introduction day, the 904 was opened up to show engine access, which was even better with back removed.

Hillclimbs were popular events for privately-entered 904s. This is Weber, German GT star, at the Schauinsland start. Frame member behind transmission served as a body mounting point, and helped protect vitals in case of a rear-end shunt during heat of battle. Sloniger photos.

Porsche opened its 904 book at Sebring in '64 but ran into clutch problems. At the Targa it all came right. When their eight-cylinder prototypes all failed it was the 904 which finished 1-3 in that Sicilian test, beating Ford but not Ferrari who had declined to play. Overall victory to a brand new GT was still remarkable.

Stuttgart insisted this was the car for tough ones. They entered a whole fleet at Nurburg Ring and despite several crashed—it was a lot of GT for private drivers—saw a GTS take 3rd overall as best Porsche. Le Mans was equally impressive: five entered, five finishing in the top dozen and the best was 7th, in fettle to go another 24 hours if asked.

Reims' 12 hour race only confirmed the 904 legend. Various private GTS were entered including one for an Argentine pair who collected their car in Stuttgart in race week, drove it all the way to central France and looked bemused when asked where their spares might be.

They slept through first training—it's a long, noisy trip from Stuttgart—hustled up a couple of borrowed lights for this night event and promptly won the 2 liter class, leading a 5-6-7 Carrera GTS sweep. After a little more sleep they drove the 904 back to Stuttgart.

Then came a 5th for the works car at high-speed Spa, a class sweep on Zolder's new track but only 2nd and 3rd to an Elan in the TT—the best-ever Porsche effort on a British airfield circuit none-the-less. In road/track events like the Tour de France where raw speed ruled Porsche ran right behind the GTO Ferraris to finish 3-4-5-6 on scratch and 1-2 on handicap.

The final proof came in January, 1965 on one of the snowiest Monte Carlo Rallies of recent memory. Porsche coaxed master chef Eugen Bohringer into a 904 for this improbable sleigh ride. Of 237 hopefuls starting, just 22 of them finished within the scoring and the doyen of German rally drivers had put his track car into 2nd overall.

Sebring that year confirmed the 904 as class champ in the distance events but a 5th in the 1965 Targa Florio also showed the limits of four cylinders—when those multis managed to finish. The car did what it was designed for—win the 2 liter class—but miracles weren't always possible. Not even with special slipperyness as proved by a win of the Le Mans Thermal Index.

By now six- and eight-cylinder works specials could leave the GT running to experienced private 904 fields. For 1966, the car was re-homologated as a sports car within the new breakdowns but the much-acclaimed GTS was never updated as Spyders had been because it was proving trickier to maintain than first expected.

Porsche not only dropped the deluxe road variant but began to ease GT racers into hot 911s when they found it was hell to replace accident-

damaged 904 panels. Also, chassis welds let go in the rough and were difficult to get at for rewelding. Dreaded corrosion broke out in their frames, hidden by the bonded-on bodies. And those bodies themselves varied widely in torsional stiffness.

When first released, the 904 did wonderfully well at burnishing the Porsche image as hoped and if it didn't enjoy the life span of a Spyder or 356 Carrera—well, racing had accelerated madly by the middle Sixties when the 904 was far more successful than most of its peers.

Add that intangible called car charisma—and the Carrera GTS was undisputed champ. It just plain looked like a racer, then went out and proved it.

Spec Charts

Engine	550-Works	Production 1954	550 1100	550A/1500 RS	Mickey Mouse	718 RSK 1958	718 RSK 1959	RSK 1600
Designer	Fuhrmann	Fuhrmann	Fuhrmann	Fuhrmann	Fuhrmann	Fuhrmann	Fuhrmann	Fuhrmann
Type	Boxer 4	Boxer 4	Boxer 4	Boxer 4	Boxer 4	Boxer 4	Boxer 4	Boxer 4
Location	Mid-engine	Mid-engine	Mid-engine	Mid-engine	Mid-engine	Mid-engine	Mid-engine	Mid-engine
Bore & Stroke, mm	85.0/66.0	85.0/66.0	73.0/66.0	85.0/66.0	85.0/66.0	85.0/66.0	85.0/66.0	87.5/66.0
Bore & Stroke, in	3.35/2.60	3.35/2.60	2.88/2.60	3.35/2.60	3.35/2.60	3.35/2.60	3.35/2.60	3.45/2.60
Displacement, cc/in	1498/91.4	1498/91.4	1098/67.0	1498/91.4	1498/91.4	1498/91.4	1498/91.4	1587/96.8
Compression Ratio	9.5:1	9.5:1		9.8:1	9.8:1	9.8:1	9.8:1	9.8:1
Camshaft Layout	D.O.H.C	D.O.H.C.	D.O.H.C.	D.O.H.C.	D.O.H.C.	D.O.H.C.	D.O.H.C.	D.O.H.C.
Camshaft Drive	Shafts & Gears	Shafts & Gears	Shafts & Gears	Shafts & Gears	Shafts & Gears	Shafts & Gears	Shafts & Gears	Shafts & Gears
Cam Followers	Finger	Finger	Finger	Finger	Finger	Finger	Finger	Finger
Valves—Design, No	Inclined-2/cyl	Inclined-2/cyl	Inclined-2/cyl	Inclined-2/cyl	Inclined-2/cyl	Inclined-2/cyl	Inclined-2/cyl	Inclined-2/cyl
Valve Springs	Coil	Coil	Coil	Coil	Coil	Coil	Coil	Coil
Sparkplugs/Cyl	2	2	2	2	2	2	2	2
Ignition—Type	2 Distributors	2 Distributors	2 Distributors	2 Distributors	2 Distributors	2 Distributors	2 Distributors	2 Distributors
Ignition—Drive	Cam-driven	Cam-driven	Cam-driven	Cam-driven	Cam-driven	Crank-driven 60°	Crank-driven 60°	Crank-driven 60°
Carburetors—No, Type	2 Solex 40 PJJ	2 Solex 40 PJJ	2 Solex 40 PJJ	2 Weber 40 DCM 1	2 Weber 40 DCM 1	2 Weber 46 IDM	2 Weber 46 IDM	2 Weber 46 IDM
BHP/RPM	110/7500	110/7000	93/5500	135/7200	135/7200	150/7800	148/8000	150/7200

Drive Train								
Clutch—Type	Single Dry-plate	Single Dry-plate	Single Dry-plate	Single Dry-plate	Single Dry-plate	Single Dry-plate	Single Dry-plate	Single Dry-plate
Clutch—Make	F&S K12 200	F&S K12 200	F&S K12 200	F&S K12 200	F&S K12 200	F&S K12 200	F&S K12 200	F&S K12 200
Transmission—Gears	4-speed & rev	4-speed & rev	4-speed & rev	5-speed & rev	5-speed & rev	5-speed & rev	5-speed & rev	5-speed & rev
Type	Porsche Synchro	Porsche Synchro	Porsche Synchro	Porsche Synchro	Porsche Synchro	Porsche Synchro	Porsche Synchro	Porsche Synchro
Location	Behind R-axle	Behind R-axle	Behind R-axle	Behind R-axle	Behind R-axle	Behind R-axle	Behind R-axle	Behind R-axle
Axle Ratios	4.375, 4.428 4.857	4.375, 4.428 4.857		4.375, 4.428 4.857	4.375, 4.428 4.857	4.375, 4.428 4.857	4.375, 4.428 4.857	4.375, 4.428 4.857

RSK 1700	RS60/61 1500	RS60/61 1600	1605 Le Mans	RS60/61 1700	RS61 1700	2000	718 F2	F2	F1
Fuhrmann	Fuhrmann	Fuhrmann	Fuhrmann	Fuhrmann	Fuhrmann	Fuhrmann	Fuhrmann	Fuhrmann	Fuhrmann
Boxer 4	Boxer 4	Boxer 4	Boxer 4	Boxer 4	Boxer 4	Boxer 4	Boxer 4	Boxer 4	Boxer 4
Mid-engine	Mid-engine	Mid-engine	Mid-engine	Mid-engine	Mid-engine	Mid-engine	Mid-engine	Mid-engine	Mid-engine
90.0/66.0	85.0/66.0	87.5/66.0	88.0/66.0	90.0/66.0	90.8/66.0	92.0/74.0	85.0/66.0	85.0/66.0	85.0/66.0
3.55/2.60	3.35/2.60	3.45/2.60	3.47/2.60	3.55/2.60	3.58/2.60	3.62/2.92	3.35/2.60	3.35/2.60	3.35/2.60
1678/102.4	1498/91.4	1587/96.8	1605/97.9	1678/102.4	1708/104.2	1966/119.9	1498/91.4	1498/91.4	1498/91.4
9.8:1	9.8:1	9.8:1	9.8:1	9.8:1	9.8:1	9.8:1	9.8:1	9.8:1	10.3:1
D.O.H.C.	D.O.H.C.	D.O.H.C.	D.O.H.C.	D.O.H.C.	D.O.H.C.	D.O.H.C.	D.O.H.C.	D.O.H.C.	D.O.H.C.
Shafts & Gears	Shafts & Gears	Shafts & Gears	Shfts & Gears	Shafts & Gears	Shafts & Gears	Shafts & Gears	Shafts & Gears	Shafts & Gears	Shafts & Gears
Finger	Finger	Finger	Finger	Finger	Finger	Finger	Finger	Finger	Finger
Inclined-2/cyl	Inclined-2/cyl	Inclined-2/cyl	Inclined-2/cyl	Inclined-2/cyl	Inclined-2/cyl	Inclined-2/cyl	Inclined-2/cyl	Inclined-2/cyl	Inclined-2/cyl
Coil	Coil	Coil	Coil	Coil	Coil	Coil	Coil	Coil	Coil
2	2	2	2	2	2	2	2	2	2
2 Distributors	2 Distributors	2 Distributors	2 Distributors	2 Distributors	2 Distributors	2 Distributors	2 Distributors	2 Distributors	2 Distributors
Crank-driven	Crank-driven	Crank-driven	Crank-driven	Crank-driven	Crank-driven	Crank-driven	Crank-driven	Crank-driven	Crank-driven
60°	60°	60°	60°	60°	60°	60°	60°	60°	60°
2 Weber	2 Weber	2 Weber	2 Weber	2 Weber	2 Weber			2 Weber	
46 IDM	46 IDM	46 IDM	46 IDM	46 IDM	46 IDM			46 IDM 1	
170/7800	150/7800	160/7800	160/7800	180/7800	180/7800	185/7800	150-165/7800	150-155/7800	180-190/8000

Single Dry-plate	Single Dry-plate	Single Dry-plate	Single Dry-plate	Single Dry-plate	Single Dry-plate	Single Dry-plate	Single Dry-plate	Single Dry-plate	Single Dry-plate
F&S K12 200	F&S K12 200	F&S K12 200	F&S K12 200	F&S K12 200	F&S K12 200	Hausserman A-12	F&S K12 200	F&S K12 200	F&S K12 200
5-speed & rev	5-speed & rev	5-speed & rev	5-speed & rev	5-speed & rev	5-speed & rev	5-speed & rev	5-speed & rev	5-speed & rev	5-speed & rev
Porsche Synchro	Porsche Synchro	Porsche Synchro	Porsche Synchro	Porsche Synchro	Porsche Synchro	Porsche Synchro	Porsche Synchro	Porsche Synchro	Porsche Synchro
Behind R-axle	Behind R-axle	Behind R-axle	Behind R-axle	Behind R-axle	Behind R-axle	Behind R-axle	Behind R-axle	Behind R-axle	Behind R-axle
4.375, 4.428 4.857	4.428	4.428	4.428	4.428	4.428	4.428	4.428	4.428	4.428

Spec Charts

	550 Works	Production 1954(2)	550 1100	550A/1500RS	Mickey Mouse	718 RSK 1958	718 RSK 1959	RSK 1600
Chassis	Welded Tube	Welded Tube	Welded Tube	Welded Tube	Welded Tube	Welded Tube	Welded Tube	Welded Tube
Type	Ladder	Ladder	Ladder	Space Frame	Space Frame	Space Frame	Space Frame	Space Frame
Wheelbase—mm/in	2100/82.7	2100/82.7	2100/82.7	2100/82.7	2000/78.8	2100/82.7	2100/82.7	2100/82.7
Track, Fr—mm/in	1290/50.8	1290/50.8	1290/50.8	1290/50.8	1150/45.3	1290/50.8	1290/50.8	1290/50.8
Track, Rear—mm/in	1250/49.3	1250/49.3	1250/49.3	1250/49.3	1150/45.3	1250/49.3	1250/49.3	1250/49.3
Suspension—Front	Independent	Independent	Independent	Independent	Independent	Independent	Independent	Independent
Type	2 Trailing Arms	2 Trailing Arms	2 Trailing Arms	2 Trailing Arms	2 Trailing Arms	2 Trailing Arms	2 Trailing Arms	2 Trailing Arms
Springs	Torsion Bar	Torsion Bar	Torsion Bar	Torsion Bar	Torsion Bar	Torsion Bar	Torsion Bar	Torsion Bar.
Suspension—Rear	Independent	Indepenent	Independent	Independent	Independent	Independent	Independent	Independent
Type	Swing Axle	Swing Axle	Swing Axle	Low Pivot Swing Axle	Low Pivot Swing Axle	Low Pivot Swing Axle	Wishbones	Wishbones
Springs	Torsion Bar	Torsion Bar	Torsion Bar	Torsion Bar	Torsion Bar	Torsion Bar	Coils	Coils
Shock Absorbers	Fichtel & Sachs	Fichtel & Sachs	Fichtel & Sachs	Fichtel & Sachs	Koni	Koni	Koni	Koni
Brakes	Drum	Drum	Drum	Drum	Drum	Drum	Drum	Drum
Wheels	Stamped Steel	Stamped Steel	Stamped Steel	Stamped Steel	Stamped Steel	Stamped Steel	Stamped Steel	Stamped Steel
Tires—Fr/Rear	5.00/5.25X16	5.00/5.25X16	5.00/5.25X16	5.00/5.25X16	5.00/5.25X16	5.00/5.25X16	5.00/5.25X16	5.00/5.25X16
General								
Body Builder	Weidenhausen	Weinsburg	Weidenhausen	Wendler	Porsche	Wendler	Wendler	Wendler
Length Overall—in	141.8	141.8	141.8	141.8	138.0	141.8	141.8	141.8
Width	60.7	60.7	60.7	61.1	59.5	59.5	59.5	59.5
Height	41.4	41.4	41.4	40.0	35.0	35.0	34.7	34.7

(2) 1954/55 chassis specs identical

RSK 1700	RS60/61 1500	RS60/61 1600	1605 Le Mans	RS60/61 1700	RS61 1700	2000	718 F2	F2	F1
Welded Tube Space Frame	Welded Tube Space Frame	Welded Tube Space Frame	Welded Tube Space Frame	Welded Tube Space Frame	Welded Tube Space Frame	Welded Tube Space Frame	Welded Tube Space Frame	Welded Tube Space Frame	Welded Tube Space Frame
2100/82.7	2200/86.7	2200/86.7	2200/86.7	2200/86.7	2200/86.7	2335/92.0	2335/92.0	2335/92.0	2300/90.6
1290/50.8	1290/50.8	1290/50.8	1290/50.8	1290/50.8	1290/50.8	1290/50.8	1300/51.2	130/51.2	1300/51.2
1250/49.3	1250/49.3	1250/49.3	1250/49.3	1250/49.3	1250/49.3	1250/49.3	1260/49.6	1260/49.6	1280/50.43
Independent	Independent	Independent	Independent	Independent	Independent	Independent	Independent	Independent	Independent
2 Trailing Arms	2 Trailing Arms	2 Trailing Arms	2 Trailing Arms	2 Trailing Arms	2 Trailing Arms	2 Trailing Arms	2 Trailing Arms	2 Trailing Arms	Wishbone
Torsion Bar	Torsion Bar	Torsion Bar	Torsion Bar	Torsion Bar	Torsion Bar	Torsion Bar	Torsion Bar	Torsion Bar	Coils
Independent Wishbones	Independent Wishbones	Independent Wishbones	Independent Wishbones	Independent Wishbones	Independent Wishbones	Independent Wishbones	Independent Wishbones	Independent Wishbones	Independent Wishbones
Coils	Coils	Coils	Coils	Coils	Coils	Coils	Coils	Coils	Coils
Koni	Koni	Koni	Koni	Koni	Koni	Koni	Koni	Koni	Koni
Drum	Drum	Drum	Drum	Drum	Drum	Drum	Drum	Drum	Ring Disc
Stamped Steel	Stamped Steel	Stamped Steel	Stamped Steel	Stamped Steel	Stamped Steel	Stamped Steel	Stamped Steel	Stamped Steel	Stamped Steel
5.00/5.25X16	5.50/5.90X15	5.50/5.90X15	5.50/5.90X15	5.50/5.90X15	5.50/5.90X15	5.50/5.90X15	5.50/5.90X15	5.50/5.90X15	5.50/5.90X15

RSK 1700	RS60/61 1500	RS60/61 1600	1605 Le Mans	RS60/61 1700	RS61 1700	2000	718 F2	F2	F1
Wendler	Wendler	Wendler	Wendler	Wendler	Wendler	Wendler	Porsche	Porsche	Porsche
141.8	145.8	145.8	145.8	145.8	145.8	145.8	141.8	132.0	134.8
59.5	59.5	59.5	59.5	59.5	59.5	59.5	59.5	36.3	33.1
34.7	38.6	38.6	38.6	38.6	38.6	38.6	34.7	35.5	31.5

Spec Charts

Engine	356 Speedster	356A 1500 GS	356A GS-GT	356A GS Deluxe	1600 GS-GT	System I	Carrera 2B	Carrera 2C
Designer	Fuhrmann	Fuhrmann	Fuhrmann	Fuhrmann	Fuhrmann	Fuhrmann	Fuhrmann	Fuhrmann
Type	Boxer 4	Boxer 4	Boxer 4	Boxer 4	Boxer 4	Boxer 4	Boxer 4	Boxer 4
Location	Behind R-Axle	Behind R-Axle	Behind R-Axle	Behind R-Axle	Behind R-axle	Behind R-axle	Behind R-axle	Behind R-axle
Bore & Stroke, mm	85.0/66.0	85.0/66.0	85.0/66.0	87.5/66.0	87.5/66.0	87.5/66.0	92.0/74.0	92.0/74.0
Bore & Stroke, in	3.35/2.60	3.35/2.60	3.35/2.60	3.45/2.60	3.45/2.60	3.45/2.60	3.62/2.92	3.62/2.92
Displacement, cc/in	1498/91.4	1498/91.4	1498/91.4	1587/96.8	1587/96.8	1587/96.8	1966/119.9	1966/119.9
Compression Ratio	9.0:1	9.0:1	9.0:1	9.5:1	9.8:1	9.8:1	9.5:1	9.5:1
Camshaft Layout	D.O.H.C.	D.O.H.C.	D.O.H.C.	D.O.H.C.	D.O.H.C.	D.O.H.C.	D.O.H.C.	D.O.H.C.
Camshaft Drive	Shafts & Gears	Shafts & Gears	Shafts & gears	Shafts & Gears	Shafts & Gears	Shafts & Gears	Shafts & Gears	Shafts & Gears
Cam Followers	Finger	Finger	Finger	Finger	Finger	Finger	Finger	Finger
Valves-Design, No	Inclined-2/cyl	Inclined-2/cyl	Inclined-2/cyl	Inclined-2/cyl	Inclined-2/cyl	Inclined-2/cyl	Inclined-2/cyl	Inclined-2/cyl
Valve Springs	Coil	Coil	Coil	Coil	Coil	Coil	Coil	Coil
Sparkplugs/Cyl	2	2	2	2	2	2	2	2
Ignition-Type	2 Distributors	2 Distributors	2 Distributors	2 Distributors	2 Distributors	2 Distributors	2 Distributors	2 Distributors
Ignition-Drive	Cam-driven	Cam-driven	Crank-driven 90°	Crank-driven 90°	Crank-driven 90°	Crank-driven 90°	Crank-driven 90°	Crank-driven 90°
Carburetors-No, Type	2 Solex 40 PJJ	2 Solex 40 PJJ	2 Solex 40 PJJ	2 Solex 40 PJJ-4	2 Weber 40 DCM 2	2 Solex 40 P-II-4	2 Solex 40 P-II-4	2 Solex 40 P-II-4
BHP/RPM	100/6200	100/6200	110/6400	105/6500	115/6500	128/6700[3]	130/6200	130/6200

Drive Train								
Clutch-Type	Single Dry-plate	Single Dry-plate	Single Dry-plate	Single Dry-Plate	Single Dry-plate	Single Dry-plate	Single Dry-plate	Single Dry-plate
Clutch-Make	F&S K12 200	F&S K12 200	Hausserman A-10	Hausserman A-10	Hausserman A-10	Hausserman A-10	Hausserman A-10	Hausserman A-12
Transmission-Gears	4-speed & rev	4-speed & rev	4-speed & rev	4-speed & rev	4-speed & rev	4-speed & rev	4-speed & rev	4-speed & rev
Type	Porsche Synchro	Porsche Synchro	Porsche Synchro	Porsche Synchro	Porsche Synchro	Porsche Synchro	Porsche Synchro	Porsche Synchro
Location	Ahead of R-axle	Ahead of R-axle	Ahead of R-axle	Ahead of R-axle	Ahead of R-axle	Ahead of R-axle	Ahead of R-axle	Ahead of R-axle
Axle Ratios	4.428	4.428	4.428	4.428	4.428	4.428	4.428	4.428

[3] System II 135/7400

C-GT	Abarth GTL	I	II	Retrofit 2.0	2000 GS-GT	904 GTS [Road]	904 GTS [Race]
Fuhrmann	Fuhrmann	Fuhrmann	Fuhrmann	Fuhrmann	Fuhrmann	Fuhrmann	Fuhrmann
Boxer 4	Boxer 4	Boxer 4	Boxer 4	Boxer 4	Boxer 4	Boxer 4	Boxer 4
Behind R-axle	Behind R-axle	Behind R-axle	Behind R-axle	Behind R-axle	Behind R-axle	Mid-engine	Mid-engine
92.0/74.0	87.5/66.0	87.5/66.0	87.5/66.0	92.0/74.0	92.0/74.0	92.0/74.0	92.0/74.0
3.62/2.92	3.45/2.60	3.45/2.60	3.45/2.60	3.62/2.92	3.62/2.92	3.62/2.92	3.62/2.92
1966/119.9	1587/96.8	1587/96.8	1587/96.8	1966/119.9	1966/119.9	1966/119.9	1966/119.9
9.5:1	9.8:1	9.8:1	9.8:1	9.8:1	9.8:1	9.8:1	9.8:1
D.O.H.C.	D.O.H.C.	D.O.H.C.	D.O.H.C.	D.O.H.C.	D.O.H.C.	D.O.H.C.	D.O.H.C.
Shafts & Gears	Shafts & Gears	Shafts & Gears	Shafts & Gears	Shafts & Gears	Shafts & Gears	Shafts & Gears	Shafts & Gears
Finger	Finger	Finger	Finger	Finger	Finger	Finger	Finger
Inclined-2/cyl	Inclined-2/cyl	Inclined-2/cyl	Inclined-2/cyl	Inclined-2/cyl	Inclined-2/cyl	Inclined-2/cyl	
Coil	Coil	Coil	Coil	Coil	Coil	Coil	Coil
2	2	2	2	2	2	2	2
2 Distributors	2 Distributors	2 Distributors	2 Distributors	2 Distributors	2 Distributors	2 Distributors	2 Distributors
Crank-driven	Crank-driven	Crank-driven	Crank-driven	Crank-driven	Crank-driven	Crank-driven	Crank-driven
90°	90°	90°	90°	90°	90°	90°	90°
2 Weber	2 Weber	2 Weber	2 Solex	2 Solex	2 Weber	2 Weber[1]	2 Weber
46 IDM 2	40 DCM 2	40 DCM 2	44 PII-4	40 PII-4	46 IDM	46 IDM	46 IDM
160/	115/6500	128/6700	135/7400		155-185/7800	155/6400	180/7000

Single Dry-plate	Single Dry-plate	Single Dry-plate	Single Dry-plate	Single Dry-plate	Single Dry-plate	Single Dry-plate	Single Dry-plate
Hausserman A-12	Hausserman A-12	Hausserman A-12	Hausserman A-12	Hausserman A-12	Hausserman A-12	Hausserman A-12	Hausserman A-12
4-speed & rev	4-speed & rev	4-speed & rev	4-speed & rev	4-speed & rev	4-speed & rev	5-speed & rev	5-speed & rev
Porsche Synchro	Porsche Synchro	Porsche Synchro	Porsche Synchro	Porsche Synchro	Porsche Synchro	Porsche Synchro	Porsche Synchro
Ahead of R-axle	Ahead of R-axle	Ahead of R-axle	Ahead of R-axle	Ahead of R-axle	Ahead of R-axle	Behind R-Axle	Behind R-axle
4.428	4.428	4.428	4.428	4.428	4.428	4.428	4.428

[1] Also Solex 44 PII-4

Spec Charts

	356 Speedster	356A 1500 GS	356A GS-GT	356A GS Deluxe	1600 GS-GT	System I	Carrera 2B	Carrera 2C
Chassis								
Type	Platform Unit/Body	Platform Unit/Body	Platform Unit/Body	Platform Unit/Body	Platform Unit/Body	Platform Unit/Body	Platform Unit/Body	Platform Unit/Body
Whelbase—mm/in	2100/82.7	2100/82.7	2100/82.7	2100/82.7	2100/82.7	2100/82.7	2100/82.7	2100/82.7
Track, Fr—mm/in	1290/50.8	1306/51.5	1306/51.5	1306/51.5	1306/51.5	1306/51.5	1306/51.5	1306/51.5
Track, Rear—mm/in	1250/49.3	1272/50.1	1272/50.1	1272/50.1	1272/50.1	1272/50.1	1272/50.1	1272/50.1
Suspension—Front	Independent	Independent	Independent	Independent	Independent	Independent	Independent	Independent
Type	2 Trailing Arms	2 Trailing Arms	2 Trailing Arms	2 Trailing Arms	2 Trailing Arms	2 Trailing Arms	2 Trailing Arms	2 Trailing Arms
Spring	Torsion Bar	Torsion Bar	Torsion Bar	Torsion Bar	Torsion Bar	Torsion Bar	Torsion Bar	Torsion Bar
Suspension—Rear	Independent	Independent	Independent	Independent	Independent	Independent	Independent	Independent
Type	Swing Axle	Swing Axle	Swing Axle	Swing Axle	Swing Axle	Swing Axle	Swing Axle	Swing Axle
Springs	Torsion Bar	Torsion Bar	Torsion Bar	Torsion Bar	Torsion Bar	Torsion Bar	Torsion Bar	Torsion Bar
Shock Absorbers	Koni	Koni	Koni	Koni	Koni	Koni	Koni	Koni
Brakes	Drum	Drum	Drum	Drum	Drum	Drum	Drum***	Ring Disc
Wheels	Stamped Steel	Stamped Steel	Stamped Steel	Stamped Steel	Stamped Steel	Stamped Steel	Stamped Steel	Stamped Steel
Tires—Fr/Rear	5.00/5.00x16	5.90/5.90x16	5.90/5.90/16	5.90/5.90X16	5.90/5.90x15	5.90/5.90X15	165X15	165X15
General								
Body Builder	Reutter*	Reutter	Reutter	Reutter	Reutter	Reutter	Reutter	Reutter
Length Overall In	155.6	155.6	155.6	155.6	155.6	156.8	155.6	156.8
Width	65.4	65.8	65.8	65.8	65.8	65.8	65.8	65.8
Height	48.1	51.6**	51.6	51.6	51.6	52.0	52.0	52.4

*Open cars by Drauz, D'Iteren   **Speedster 48.1   *** or Ring Disc

	C-GT	Abarth GTL	I	II	Retrofit 2.0	2000 GS-GT	904 GTS [Road]	904 GTS [Race]
	Platform	Platform	Platform	Platform	Platform	Platform	Welded Steel	Welded Steel
	Unit/Body	Unit/Body	Unit/Body	Unit/Body	Unit/Body	Unit/Body	Box Members	Box Members
	2100/82.7	2100/82.7	2100/82.7	2100/82.7	2100/82.7	2100/82.7	2300/90.6	2300/90.6
	1306/51.5	1306/51.5	1306/51.5	1306/51.5	1306/51.5	1306/51.5	1314/51.8	1314/51.8
	1272/50.1	1272/50.1	1272/50.1	1272/50.1	1272/50.1	1272/50.1	1312/51.7	1312/51.7
	Independent	Independent	Independent	Independent	Independent	Independent	Independent	Independent
	2 Trailing Arms	2 Trailing Arms	2 Trailing Arms	2 Trailing Arms	2 Trailing Arms	2 Trailing Arms	Wishbone	Wishbone
	Torsion Bar	Torsion Bar	Torsion Bar	Torsion Bar	Torsion Bar	Torsion Bar	Coil	Coil
	Independent	Independent	Independent	Independent	Independent	Independent	Independent	Independent
	Swing Axle	Swing Axle	Swing Axle	Swing Axle	Swing Axle	Swing Axle	Wishbone	Wishbone
	Torsion Bar	Torsion Bar	Torsion Bar	Torsion Bar	Torsion Bar	Torsion Bar	Coil	Coil
	Koni	Koni	Koni	Koni	Koni	Koni	Koni	Koni
	Ring Disc	Drum: Retrofit Disc →				Ring Disc	Disc	Disc
	Stamped Steel	Stamped Steel	Stamped Steel	Stamped Steel	Stamped Steel	Stamped Steel	Stamped Steel	Stamped Steel
	165X15	5.90X15	5.90X15	5.90X15	5.90X15	5.90X15	5.50/6.00X15	5.50/6.00X15

	Reutter	Abarth	Abarth	Abarth	Abarth	Porsche	Heinkel	Heinkel
	156.8	156.8	156.8	156.8	156.8	158.0	161.2	161.2
	65.8	65.8	65.8	65.8	65.8	61.2	60.7	60.7
	52.4	52.0	52.0	52.0	52.0	46.5	42.0	42.0

Specials

Maria de Filippis attempted to qualify Jean Behra's barely-finished, Colotti-built Porsche single seater (No. 4) at Monaco in 1959, without success. Molter photos. Finally, in French racing blue, Herrmann drove the car to 2nd at Reims, and led the factory Formula 2 car home. Coltrin photo.

A Porsche special at the Targa Florio. Goddard photo.

Tim Meyer's Carrera special, at the Nurburg Ring in 1963. Body builder unknown.

Hans Herrmann in a modified 550A/1500RS. Below, rare as Mickey Mouse, the specially-bodied Storez (French race/rally driver) Spyder built for France and destroyed on its first outing in 1958. Weitmann photos.

ACKNOWLEDGEMENTS

All books are the result of the combined efforts of many persons. This book could not have been published without the expert help of author Jerry Sloniger, technical assistance of Jim Wellington and Dave Love and the following:
BOOK DESIGN—Chuck Queener
COVER COLOR SEPARATIONS—Capper, Inc., Knoxville, TN
COVER PHOTOS—Chuck Queener, Jerry Sloniger,
    Bob Tronolone
TYPESETTING—Haessner Publishing, Inc, Newfoundland, NJ